Rebuilding and Tuning

Ford's Kent Crossflow Engine

Peter & Valerie Wallage

Haynes®

First published 1995

Reprinted 1996

Published by:
Haynes Publishing
Sparkford, Nr Yeovil, Somerset
BA22 7JJ, England

British Library Cataloguing in Publication Data

A catalogue record for this book is available from the British Library

ISBN 1 85010 938 9

Library of Congress catalogue card no. 94-77767

Printed in Great Britain by J. H. Haynes & Co. Ltd.

While every effort is taken to ensure the accuracy of the information given in this book, no liability can be accepted by the author or publishers for any loss, damage or injury caused by errors in, or omissions from, the information given.

Cover illustrations:
Front *Courtesy Ford Motor Company*
Back *Courtesy* ***Which Kit?*** *magazine*

Contents

1	The scope for tuning the Crossflow	5
2	Rebuilding and tuning in your home garage	11
3	Getting more power: the general principles of tuning	23
4	The intake side	40
5	The exhaust side	51
6	The ignition system	60
7	Stripping the engine	74
8	The cylinder head	95
9	The valve gear	105
10	Camshafts	109
11	The crankshaft and flywheel	119
12	The block and crankcase	125
13	Pistons and connecting rods	130
14	The lubrication system	136
15	Rebuilding the engine	140
16	Overhauling a Weber DGAV 32/36	169
	Useful information	184
	Appendix: Specifications & Data	185

Acknowledgements

Our thanks are due to the following companies for help and advice in preparing this book and for giving us facilities in their workshops for taking photographs.

Burton Performance Centre Ltd.
Oselli Engineering Ltd.
Specialised Engines Ltd.
Vulcan Engineering Ltd.
HT Racing Ltd.
Piper RS Ltd.
Kent Performance Cams Ltd.

The following companies also gave useful help and advice:

Ford Motor Company Ltd.
Pirelli Ltd.
K&N Filters (Europe) Ltd.
Weber Concessionaires Ltd.
Autocar Equipment Ltd.
Janspeed Engineering Ltd.
Induction Technology Group Ltd.

Photographs and illustrations:

Valerie Wallage
John Wallage
Ford Motor Company
K&N Filters (Europe) Ltd.
Induction Technology Group Ltd.
Pirelli Ltd.
Oselli Engineering Ltd.
Kent Performance Cams Ltd.
Piper RS Ltd.
Which Kit? magazine.

Chapter 1

The scope for tuning the Crossflow

The Crossflow Kent engine, so named because the chief engine designer, Alan Worters, lived in Kent, was one of the most successful engines Ford ever made. It first appeared in September 1967 in the Mk II 1300 and 1600 Cortinas, and carried on into the Mk III Cortinas and the 1300 Mk IV. It also powered the 1300 and 1600 Mk I Capris and the 1300 Mk II and Mk III models, the 1100, 1300 and 1600 Mk I and Mk II Escorts and lived on in transverse form in the XR2 Fiesta.

It's also been used as a popular engine option in a number of small production sports cars like Caterham, Westfield and Morgan, and in numerous kit car designs which use a Ford as the donor car. There have been different capacities for overseas markets, and thousands of Kent engines have been built for industrial use – powering everything from compressors to dump trucks.

Ford itself does not have accurate figures available for the total production of the Kent engine in all its forms – vehicle and industrial – but gives an educated guess at 'somewhere approaching 10 million'.

In motor sport the Kent achieved lasting fame as the engine which drove the Ford team to victory in the 1970 Mexico World Cup rally. It was the backbone of Formula Ford racing cars and countless privately-entered sports and saloon racers tuned, on occasions, to produce almost indecent power.

With its very strong five-bearing bottom end it has always been, and still is, a favourite for tuning, both by specialist firms and do-it-yourselfers in their home garages.

There have been other Ford engines with crossflow cylinder heads. The Pinto and CVH, for example, are crossflow designs, but to Ford enthusiasts there is only one Crossflow with a capital C, the long-lived and still immensely popular Kent engine. Because most enthusiasts refer to it as the Crossflow rather than the Kent, I shall as well.

Basically it is an engine which is very simple and rugged – qualities which often lend themselves to tuning for high power, and it's an ideal DIY engine. One of its best qualities from the home mechanic's point of view is that you can do just about everything on it with a relatively modest tool kit. You can, if you want, go out and spend a fortune on tools – including nearly 60 special tools listed by Ford – but there's really no need. Some of these tools are designed purely to make the job quicker, and you can do without them if you aren't counting the hours. Others, such as valve spring compressors and piston ring clamps, can be replaced by 'universal' tools from your local accessory shop. A few, like valve seat cutters, are essential but I would suggest that machining work like this is best put out to specialist engine firms.

The Crossflow was considerably strengthened in mid-1970 with a stiffer crankcase, stronger main-bearing caps, larger diameter cam followers and various other modifications. The stronger engines are the ones to go for if you have higher power in mind. They are commonly known as 'Seven-eleven M' engines because the casting number, which is cast in raised letters on the side of the crankcase, starts with 711M. If you want to stay with the earlier engine for any reason – either because it's the one you've got or you need it to keep the car in period as a classic – fine. The pre-711M engines can be tuned just the same as the later ones but, if you have the choice, go for the stronger 711M.

It is possible to take the peak power of the Crossflow up to more than double the standard Ford horsepower, but here we are talking about engines built for racing where the power comes in over a very narrow high rev range. This is state of the art tuning and it goes way beyond what the vast majority of enthusiasts want or need. For a start, it is hideously expensive as nearly all the moving parts have to be replaced with special components, and the engine becomes so lacking in tractability that it is almost undriveable with any finesse below about 3000 or 4000 rpm.

A stage below this is what is often loosely termed 'a rally-tune engine' and it is certainly more tractable than a racing engine. Even so, few engines of this output are tuned in a home garage and then only by experienced engine builders.

The majority of home tuners aim for what is generally called 'a fast-road engine' with outputs anything from 10 to 30 bhp more than Ford's standard tune.

All tuning is a compromise, and the higher you go the fewer compromises you can afford to make. You have to decide what you want to use the car for. If it's purely for competition work and the car is carried to and from events on a trailer it doesn't really matter if it's impossible to drive in traffic without constantly slipping the clutch to keep the revs up, or having to keep blipping the throttle while waiting at traffic lights because at tickover, even a fast tickover, the engine shakes itself like a dog coming out of a river.

If, at the other end of the scale, you want to use the car for going to work, shopping, taking the kids to school and generally what the insurance companies call social and domestic use, you will have to sacrifice top power for good slow-running and tractability. Here, your tuning would be aimed at increasing the power in the middle rev ranges, giving you lively response, good acceleration and effortless motorway cruising. It's

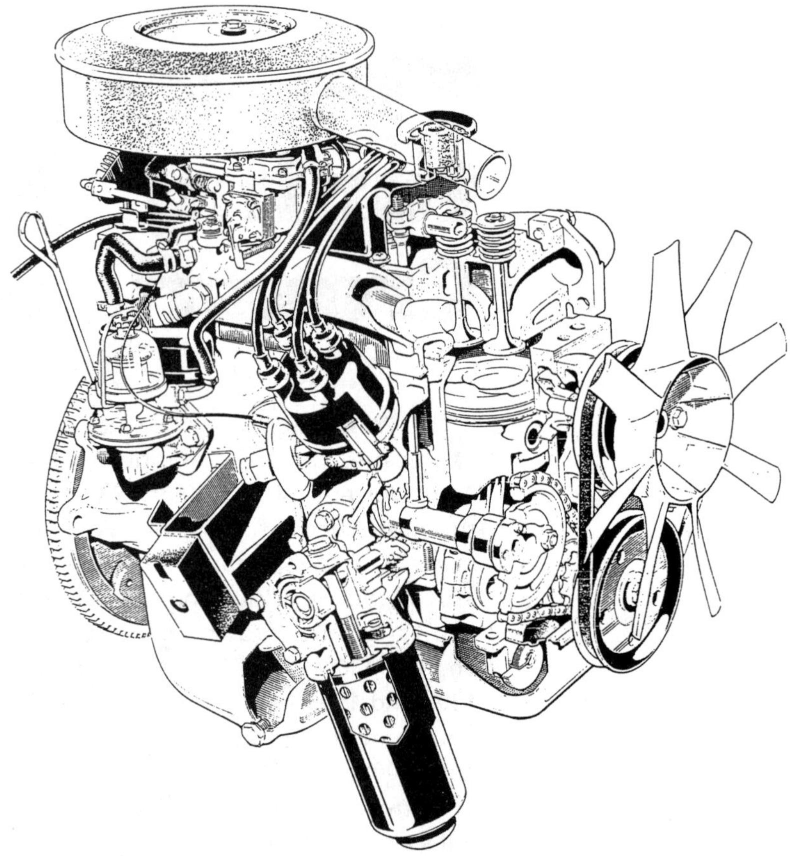

▲ *In 18 years of production, nearly 10 million Kent Crossflow engines were produced in capacities from 1100cc to 1600cc.*

The first cars to use the Crossflow were the Mk II Cortinas in September 1967.

Early Mk I Escorts used the pre-711M engine on which the crankcase and main-bearing caps were not quite so robust as the later engines. These engines are still very tuneable for road use, but for higher power go for the 711M introduced in mid-1970.

All Mk II Escorts used the 711M engine. This is a 1976 1300.

very much a question of horses for courses, and you don't buy a racehorse to put between shafts.

There are two nice things about the Crossflow. The first is that it was made in such enormous quantities that there are plenty of them about. At the time of writing, almost every breaker's yard has at least two or three examples at low prices. As a check on this, I called in at my local friendly breaker and was offered a choice of three 1600 Crossflows. One, 'It was running when we took it out, Guv, so we want forty-five quid for it', was standing washed off ready for collecting. The other two were still in cars, 'Take them out yourself, Squire, as seen, twenty-five quid apiece.' At these sorts of prices you can afford to buy one for tuning, and work on it while your car is still on the road.

The second nice thing about it is that, provided it is in good condition and not a worn-out, oil-burning old banger with compressions all over the place, you can start tuning to a mild pep-up for a few pounds with the engine still in the car, and progress from there as funds and opportunity allow.

A system followed by some enthusiasts is to tune the engine in the car to the extent where it has to come out for attention to the bottom end. Then they buy a second engine, have the bottom end attended to and rebuild it using the tuned parts from the car's engine.

There comes a point where increased power from the top end is going to put such extra stress on the bottom end of the engine that even the rugged five-bearing bottom end of a Crossflow opts out, sometimes quite spectacularly. The youngest Crossflow is over 10 years old and the majority far older. As they get older, in what are basically cheap second-hand cars, servicing and oil changes are neglected and, though the engine might be described as 'a good runner, very reliable', it is likely to be too tired to take very much by way of increased power output.

For anything more than a very modest tuning I would strongly recommend that

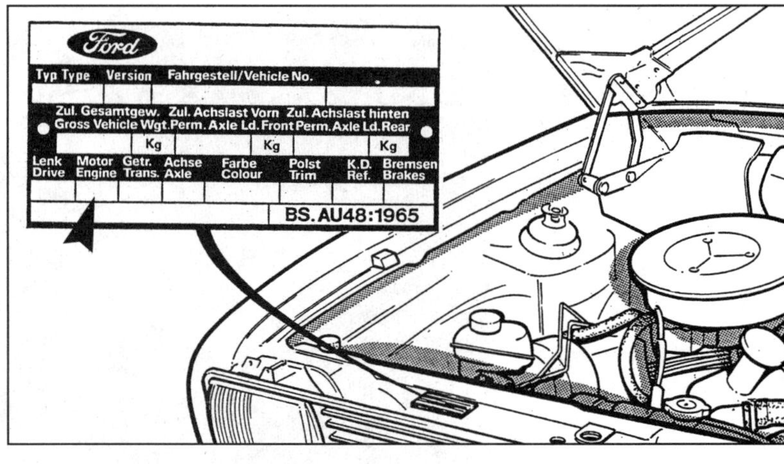

▲ *If you're looking for a second engine to rebuild, you can identify its type from the car's VIN plate. This is where you find the engine type on a Mk II Escort: **G1** indicates 1100cc low compression,**G2** is 1100cc high compression, **J1** is 1300cc low compression, **J2** is 1300cc high compression, **J3** is 1300cc fitted with a twin-choke carburettor and **L3** is 1600cc. There was an 1100cc with a twin-choke carburettor, code **G3**, but this was for Italy and France only.*

▼ *On a Mk I Escort, the engine code is second from the right at the top of the VIN plate. **B** indicates 1100cc high compression, **C** is 1100cc low compression, **S** is 1300cc high compression,**T** is 1300cc low compression and **R** is 1300 GT.*

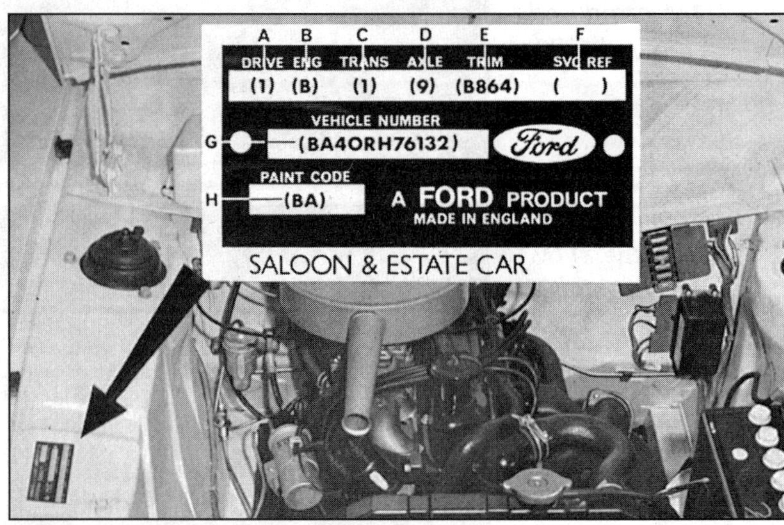

▶ *Even when the Weber twin-choke carburettor is fitted as standard, the large Ford air filter restricts the air flow.*

▶ *An easy first step to tuning is to replace the standard Ford filter with a free-flow filter such as a K&N. This engine also has a performance exhaust manifold. The large tubular bar across the top is a suspension brace.*

▼ *Ford's standard exhaust system also strangles the engine and fast-fit types are no better. You will get extra power with a proprietary performance exhaust system.*

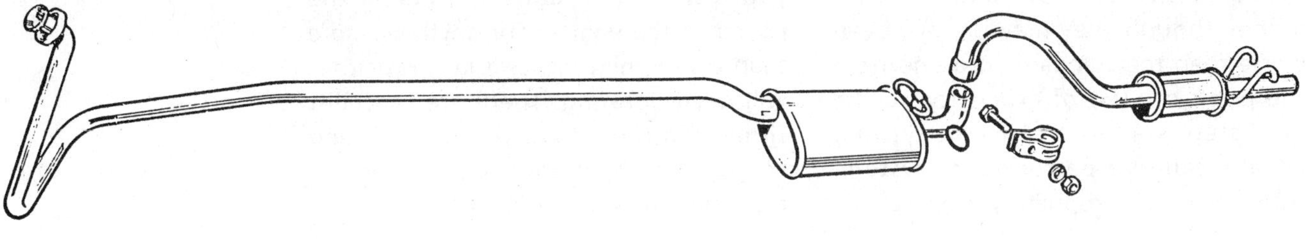

the engine is stripped and overhauled. The further you intend to take the tuning, the more fussy you should be about making the bottom end bullet proof.

Let's look at a few examples. Starting with just changing components, you can give the Crossflow a useful amount of extra response, and probably better economy into the bargain, by overhauling the ignition system either with a rebuilt or a higher performance distributor, a decent set of leads and plugs and a free-flow air filter. This alone will give you four or five more bhp. It doesn't sound a lot, but the better response will make it feel more than that. You might have to change the air filter again if you decide to swap carburettors later on, but the distributor will stand you in good stead for a lot more power to come.

So far you've improved the inlet breathing and given the engine the means to burn the extra mixture properly. The next stage is better scavenging, getting rid of the burnt exhaust gases. Ford's standard exhaust manifold for the Crossflow ranged from quite good on the GT and Sport versions to downright horrible on some of the basic saloons. If you've got one of these you can improve things a lot by changing to one of the better manifolds. Once again, your local friendly breaker is an invaluable source of supply. Ford's standard exhaust system for the Crossflow from the manifold backwards was aimed much more at quiet refined running than efficiency but I doubt if many, if any at all, Crossflows are running with their original system. Most of the cheap fast-fit systems, value for money though they may be, are even worse than the standard Ford one from the point of view of efficiency, so the next step is a more efficient system, usually called a performance system, from one of the proprietary makers.

Unless you already have a Sport or GT version, the next bottleneck in power production to be changed is the single choke carburettor. Once again, you can use standard Ford parts by getting an inlet manifold from a 1600 GT or Sport, complete with its Weber twin-choke carburettor, or you can get the same carburettor with different jets from a Pinto-engined Ford. You'll probably have to change the jets, but your local Weber agent will be able to advise you.

Provided your engine was in good condition to start with, you will by now have a car that is transformed from its original family saloon performance. After that you start moving on to gasflowed heads, larger valves, higher lift camshafts, balanced cranks and pistons, and so on, and it's time to have the engine out for a complete rebuild.

Before we go on to deal with tuning and rebuilding in detail, a couple of words of warning. First, however far you go in your tuning, you must let your insurance company know what you've done, and preferably what you intend to do. If you don't, then, in the unhappy event of an accident and a visit from the insurance company's engineer, you will find that your insurance company will be very upset. In an extreme case they may declare your policy void because you have broken its terms, and you might find it very difficult to get insured again. In any event, you will be faced with a very heavy increase in premium, probably far heavier than if you had told them about the power modifications in the first place.

Second, the brakes and suspension on your car were designed to match the power of the engine. For anything more than a very mild increase in power they will need uprating. There are specialist firms who can advise you on this, and you can find their advertisements in the magazines which cater for tuned cars.

Chapter 2

Rebuilding and tuning in your home garage

The very fact that you're reading this book presupposes that you are going to do much of the work on the engine yourself. You're fired with enthusiasm and looking forward to the day when you sit behind the wheel of your car with a sweetly running, powerful engine under the bonnet.

Enthusiasm is a fine thing. Without it little would get done, but we have to temper enthusiasm with a little practicality. It's far better, after a few months, to be on the road with a lively, reliable and tractable engine that's a pleasure to drive, than standing in a garage with boxes full of unfinished components, even if they have all the potential in the world, and little chance of ever getting them together again as a running engine.

Quite a few home-tuning projects are abandoned, usually because of lack of cash, lack of skill, lack of equipment or lack of knowledge. I hope that this book will give you the knowledge you need. Only you can decide on your resources of cash, equipment and skill.

Try to cost the project as accurately as you can before you start. To some extent the cost will depend on the condition of the engine, but it will be governed by how much extra power you want from it. You might be lucky to

▶ *Try to cost your tuning before you start. A good tuning firm should be able to give you costs for various stages from a mild pep-up to full racing tune.*

◀ *When you are reconditioning for fast road work you'll need most of the components shown here either new or reconditioned.*

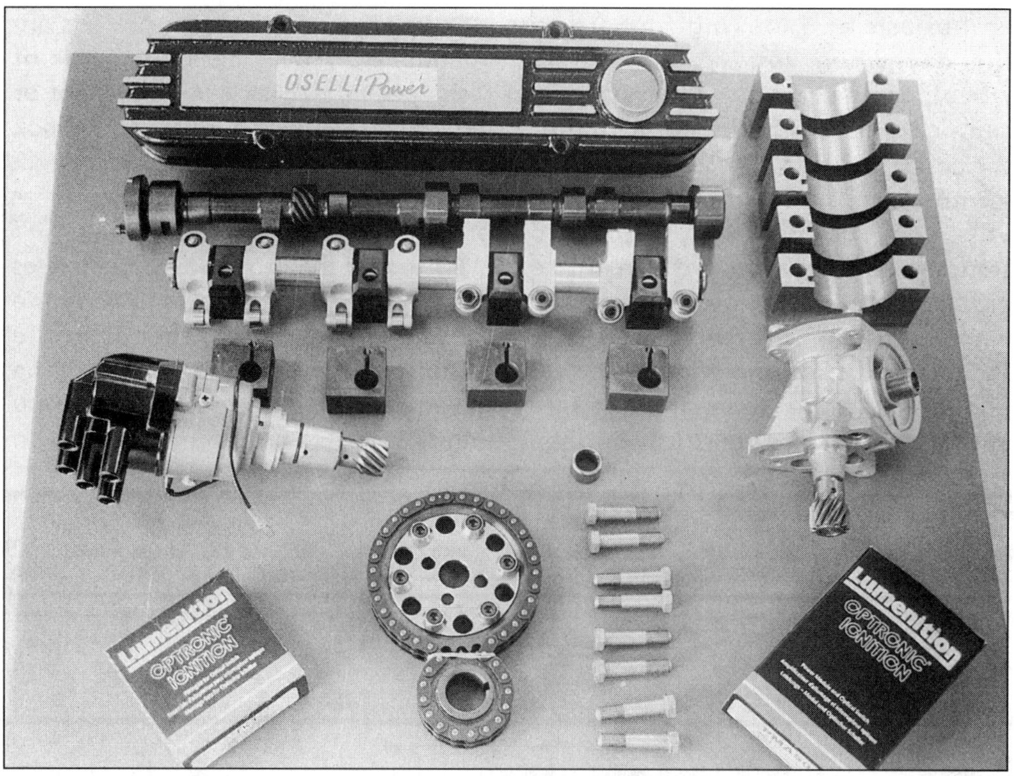

◀ *When you go for higher power you move into the realms of steel bearing caps, roller rocker shafts and other expensive items.*

have, or to find, an engine that has relatively little wear so that you can give it a new lease on life and a mild pep-up to make it more lively and responsive without reboring the cylinders, buying new pistons and regrinding the crankshaft.

At the other end of the scale you might be landed with a real dog of an engine with the bores badly worn, the crank journals down to their minimum diameter and the rest of the components so worn that they're fit only for scrap. All you could reasonably expect to salvage would be the head casting and perhaps the block, with the likely need of having to put liners in the bores. You would be better off looking for another engine.

▶ If you discover that your crankshaft is ridged and worn when you take the bearing caps off, you've no option but to have it reground.

▶▶ At the minimum you'll need a new timing chain, probably new chainwheels as well.

▶ If the bores show very little wear, for mild tuning you might get away with plain honing and new piston rings. This is a heavy duty honing machine, but you can get efficient honers to fit in your home electric drill.

Most engines come somewhere between these two examples and, before you start costing for extra power, you should work out the cost of reconditioning the engine back to standard. Some of these costs – like grinding the crankshaft, reboring the cylinders, new pistons, new timing chain and so on – will be about the same whether you are reconditioning for standard power or rebuilding for extra power. If you plan to use uprated components, then the cost for extra power may be slightly higher but, in general, my advice is to cost first to bring the engine back to standard and then modify this with the extra costs of tuning.

It's difficult to cost accurately before you strip the engine, but always err on the generous side. Overestimating and having some cash to spare is always a nicer feeling than underestimating and having to dig deeper.

Costing checklist

As a working basis for building a fast road engine, allow for the following, always hoping for the best but costing for the worst. A best-case scenario and a worst-case scenario, as American tuners often put it. Usually, the actual cost will lie somewhere in between. Get yourself some catalogues and price lists from tuning firms and start jotting down figures. You may not frighten yourself as much as you think.

Cylinder head face

Crossflow engines often suffer from slight warping of the head face after large mileages. If you are just rebuilding for extended life on a very tight budget you may decide that refacing the head is not warranted. If you are going just for new piston rings and bearing shells you might get away without refacing. But, if you want top reliability and, particularly, if you have extra power in mind for later tuning, any warping is too much. Refacing for a slight warp will cost just as much as refacing for a large one.

Valves and valve seats

It's very difficult to say what you will have to renew before you strip the engine because you'll have no idea of what you're going to find under the head. As a very minimum for extended life you will want the valve seats recut, and the valves refaced. Any local engineering shop, and some garages, will do this for you, but don't necessarily go for the cheapest quote.

On cut-price jobs the valve seats will probably be recut with something like a Vibro-Centric tool, rather like a large hand drill, which recuts the seats by grinding them with a shaped stone. It's fine for rebuilding to standard on a lot of engines, but Ford engines should have the valve seats recut to three angles, 20°, 45° and 60°. With stone grinding you will most likely get only the one cut at 45°. The engine will still run perfectly well, but a single cut is hopeless if you are tuning for extra power.

There is also the question of valve pocketing. The engine may already have had a cheap reconditioning, and there's no way you can tell till you get inside. If it has, and the valve seats have already been stone ground once, a second grind is going to sink them so far into the head that the valves will sit below the surface. Once again, it will run, but power output will be down. On a head where the seats have been ground a couple of times, they won't take a third without serious drop in performance.

◀ A minimum requirement will be to have the valve seats recut, and possibly the head refaced as well. Ford valve seats are machined at three angles. Cheap single-angle stone refacing will cut power drastically.

The only answer here is to have valve inserts put in to bring it back to standard, or look for a less worn head. Having valve seat inserts put in may not be such a bad idea, because you are halfway towards converting the head for unleaded fuel, though normally on a Crossflow only the exhaust valves need to have hardened inserts for unleaded. The other half of conversion is to have stainless steel valves.

If you're going for more than mild tuning you will need larger valves, or at least larger inlet valves, and this is where you forget your local general engineering shop and garage and visit a specialist tuner. You will get nothing from larger valves if you don't have the valve ports ground to suit them. This isn't just a case of getting a grinding stone and happily grinding away at it.

Quite apart from the danger of breaking through into the water jacket, the port has to be enlarged and shaped to get the best gas flow of the extra mixture that the larger inlet valve will let past the head. Experience counts, and established tuning firms have ported so many Crossflows that they know exactly what they're doing.

▶ *Ford's standard air filter is restrictive, but it can be improved by using a free flow element inside.*

▶ *You get a better air flow with a proprietary filter like this K&N.*

Valve springs

Trying to reuse old valve springs is a complete waste of money. You save pennies but nullify a lot of the other money you spend. In some older books on engine overhaul you see diagrams of comparing old and new valve spring lengths when they're compressed in a vice to see whether or not the old springs are fit for further service. Forget it. This is a hangover from the old days when compression ratios were in the order of 6 to 1 or less and 3,000 rpm was considered high. Fit new ones as a matter of course.

Whether or not you need to fit double valve springs depends on whether or not you are going for extra power. Let's get one thing about valve springs cleared up before we go any further. Tuners invariably use more than one spring on each valve, usually two but sometimes three on very highly tuned racing engines. This isn't because they want extra spring-power to hold each valve more tightly on its seat – on both the compression and the power strokes, when the valves are closed, the pressure inside the cylinder holds the valves tight on their seats.

The purpose of multiple valve springs is to ensure fast closing and to stop valve bounce. On a four-stroke engine each valve opens once every other revolution of the crankshaft. So, at (say) 6,500 rpm engine speed, each valve is lifted off its seat and closes again 3,250 times a minute, or 54 times every second. That's pretty fast, and as coil springs have a natural frequency, somewhere in the rev band (usually in the higher range) the natural frequency of the spring is going to coincide with the frequency of opening and closing the valve and this will cause the valve to bounce slightly on its seat each time it closes.

Put a second, quite light, coil spring inside the main one and the combined natural frequency of the two springs is lifted right out of the engine's rev range so the valves don't bounce.

If you do decide on double valve springs then, once again, you need to visit a specialist tuner. There are two reasons why you can't just pop an extra spring inside the standard Ford spring – the valve spring cap won't accept it or locate it and the chances are that the standard valve stem oil seal will also get in the way.

For higher power than just a mild tuning, there's another thing to be taken into account. When you have valve seats recut, and fit either new valves or refaced valves, the valves aren't going to sit in exactly the same position that they did before. They will be either higher or lower in the head. This means that the distance between the underside of the valve cap and the top face of the head, in other words the compressed length of the valve spring, will vary.

For smooth running and maximum efficiency the compressed lengths should all be the same and, moreover, the springs should be compressed sufficiently to give maximum poundage without becoming coil-bound. The term 'coil-bound' means that the spring has been compressed till the coils touch each other and it goes solid, like a piece of tubing. If the camshaft is still trying to open the valve at this point something's got to give. If you're lucky, the push rod bends. If you're unlucky, something more expensive goes.

▲ *Most makers of performance camshafts also specify double valve springs to stop valve bounce at high revs.*

◀ *With a high-lift camshaft you must have the spring heights checked to make sure the coils don't touch.*

Valve guides

New valve guides are essential on anything except a plain decoke – which is just postponing the day, not very successfully as a rule, when more work needs to be done on the head. Unlike many engines, Ford valve guides are not separate as standard; they are cast in the head. Ford used to supply valves with oversize stems and reamers for the valve guides to suit, but I haven't seen them around for a long time and, in any case, they don't really have a place in a tuning workshop.

Crossflow valve guides are vulnerable to wear because, compared with many engines, they are short. Unlike the Pinto, for example, there is little or no boss protruding down into the combustion chamber, which makes for good flow characteristics but doesn't do much towards getting long life from the guides. Any tuning shop will machine out the old cast-in guides and press in new separate ones for you. Indeed, it's essential to have this done if you're having the valve seats recut, because the mandrel guide for the cutter runs in the valve guide and, if this is worn, you won't get a truly concentric cut.

Rockers and rocker shaft

The best advice here is to fit new ones. The old rockers and shaft will be worn, with the pads on the valve ends of the rockers indented. If you fit them again you'll end up with a noisy engine and you'll never get your timing spot-on.

Cylinder block

On the Crossflow this is a pretty substantial lump of metal, and it's rare to find the face badly distorted, but check it just in case. If you are going for anything more than a mild tune, a light skim may be advisable, but be guided by the firm that is reboring the cylinders.

For anything more than just an overhaul you will be having the cylinders rebored, and the degree of oversize you choose may be dictated by previous overboring. If it isn't, you have the choice of boring just enough to clear the wear, or going to the maximum to get more capacity. The maximum safe overbore is plus 0.090 in, which will increase the capacity of the 1600 engine to what is commonly called a 1700, though the exact figure is 1691cc.

The bottom end of a Crossflow rarely gives trouble, though after very high mileage the bores for the main bearings can go slightly oval. On a 711M engine it's very rarely anything to worry about unless you are going for top power, then it is advisable to allow in your costing for having the bores checked and, if necessary, have the caps let down slightly and the housings line-bored and reamed.

Crankshaft

Allow for this to be reground and polished, with new mains and big-end bearings and thrust washers. There is a lot of loose talk about the necessity for having the crankshaft Tuftrided. This doesn't do a great deal for crankshaft strength but it gives the journals a very much harder face. The process involves bringing the crankshaft up to a high temperature in a chemical bath. Unfortunately, this can cause distortion, and Tuftrided crankshafts often have to be straightened, so the whole process can be expensive. If you're going for out-and-out power in competition, then it's worth the extra cost, but for road use I would think long and hard about it and probably come down on the side of saving my money to spend on other parts of the engine, or on the brakes and suspension.

Balancing is a different matter, and

you should allow for having the crankshaft, flywheel and clutch balanced. The clutch side of the flywheel will probably need surface grinding to take out the ridges and, for tuning, I would advise having it lightened by machining the crankshaft side as well.

Ford fitted a number of different weights of flywheel to the Crossflow, and the heavier ones can with advantage be lightened to give you better response and acceleration. Don't go silly on it, though, or you'll end up with an engine that has vivid acceleration but very rough and lumpy slow-running. I would advise 13 to 14 lb as a minimum weight even on a highly-tuned road engine. Allow for having a new starter ring gear fitted.

Connecting rods

These should be checked for alignment, for ovality of the big-end housings and for wear in the small-end bush. This can be replaced quite cheaply but, if the rods are distorted, check carefully the price of correction or new ones. Ford no longer does an exchange on connecting rods – only outright purchase – and you can save money by looking for undamaged second-hand ones. On a tuned engine, above anything mild, allow for having the rods and pistons balanced.

Camshaft

When you look through catalogues you will be faced with a bewildering choice of performance camshafts for the Crossflow, and I will talk about these more fully in a later chapter. You might have to have the camshaft bearings replaced, or possibly just the front one, but it's not all that usual.

You will need new cam followers, a new timing chain and, for tuning, go for an adjustable (sometimes called vernier) chain wheel. For top power, a duplex timing chain gives peace of mind, but for normal fast road work a single chain is quite adequate. You can't fit a duplex chain on the Fiesta Crossflow, and you can't fit the usual type of vernier chainwheel with a ring of bolts because the timing cover is too shallow. Some specialist firms, Vulcan Engineering for example, offer a special vernier chainwheel for Fiestas.

▲ *Higher stages of tuning involve head porting and larger valves.*

Oil pump

Ford used two types of oil pump on the Crossflow, a rotary vane type and an eccentric rotor type. The rotor type is later and better. In most manuals you will find instructions for rebuilding this but, as the oil pump is the heart which pumps the engine's life blood, it's not worthwhile chancing reconditioning it. If yours is worn, fit a new one and, for tuning, it's worthwhile fitting a higher pressure one than the standard or modifying yours for higher pressure if it's unworn. For top power you can go for one which also has a higher capacity, but for fast road use the extra cost is hardly justified.

Extras

New gaskets, oil seals, core plugs, quite a few new bolts, new lock washers and locking plates, having the parts you are reusing thoroughly cleaned and scoured in a hot chemical bath.

It's quite a formidable list and, as most of the reconditioning involves machining or the use of special tools, you will have to have a lot of the work done outside. Where you will save over the cost of buying a reconditioned engine is the labour.

A word of warning here. Don't be tempted to buy a 'cheap' reconditioned or rebuilt engine as a basis to start tuning. Reconditioning an engine properly is not cheap. The only way anyone can cut the price is to widen the tolerance of wear on the basis that 'it's worn, but it'll last a good few thousand miles', and by employing labour on piece-work where speed is more important than careful fitting (leaving things to bed down during use – they will, but with built-in wear!). Such engines might last quite well under local shopping-trip conditions, but under the stresses of increased power the cheapening could show up as rapid wear or even failure within a few hundred miles.

You have to modify and add to the cost of the reconditioning list depending on how much extra power you want. On the cylinder head you should add gas-flowing of the inlet and exhaust ports, and possibly larger valves depending on the degree of tune. If you go for a high-performance camshaft with a high lift, you may have to have the tops of the pistons machined to clear the valves, and so on. I will deal with all these in later chapters.

You will be wasting much of the money you spend on head tuning if you neglect a performance exhaust system, a free-flowing air filter, and possibly a change of carburettor depending on the one already fitted.

If you are working on a strict, and restricting, budget and are not looking for top power, it's possible to cut down on this cost considerably and still finish up with a reliable engine giving more power than the one in standard tune. The big proviso is that you start with an engine which is in good condition and, I would suggest, preferably with no more than 60,000 to 65,000 miles on the clock. Some higher mileage engines may still be in good condition but, generally, the higher the mileage the more you have to do to recondition the engine.

Where you save most money is on the bottom end. The Crossflow engine is pretty robust downstairs and, if it's not badly worn, you can give it quite a few more thousand miles with a change of bearing shells, new timing chain and tensioner, new piston rings and honing the bores to break the glaze.

At the top, you will want a minimum of valve refacing, new valve guides, new valve springs and an overhaul of the rocker gear.

If the engine is in good condition, you can tackle the early stages of tuning without taking the engine out of the car.

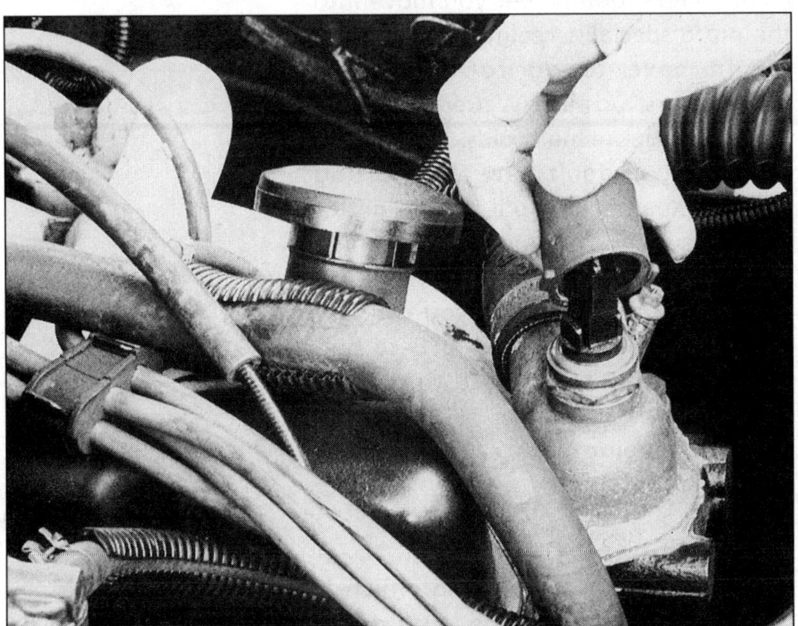

▼ *A simple modification of using the thermostat housing from a later Ford gives you a thermostatic switch to control an electric fan. This gives you the extra 2 to 3 bhp that the belt-driven fan consumed.*

I will assume you have a running car with a Crossflow engine, and I'll look at the various stages through which you can take the tune. The first stages can be carried out without removing the engine from the car.

The biggest restricting factor in engine power is the breathing, so it makes sense to tackle that first. Looking at the inlet side, you will need a free-flowing air filter such as a K&N or ITG Megaflow and, possibly, a different carburettor depending on the one already fitted. You should make your mind up about the carburettor before you buy a new air filter as the fixings will differ accordingly.

On the smaller capacity Crossflows, the single-choke Weber, usually an ICH, won't pass enough mixture to give you much in the way of extra power. Ford's Motorcraft is a good workmanlike carburettor, but not really suited to a tuned engine.

Ford's VV (Variable Venturi) carburettor is one which has a mixed reputation with tuners. Some like it and some won't have it at any price. When it's working well on full song it's a lovely carburettor and hard to beat, at moderate power increases, for combining response with economy. Unfortunately, it doesn't seem to hold its tune as well as Weber carburettors, and when it goes off song it can be a very troublesome carburettor to strip and retune at home. If you want to stay with it, find someone with considerable experience of the beast and get them to set it up for you.

Most tuners go for a double-choke Weber carburettor and, on the 1600cc engine, you might possibly have one already fitted. If not, you have the choice of buying a new one or buying a second-hand one, and a manifold to suit, and reconditioning it. Weber carburettors look complicated but they are quite straightforward and not very difficult to overhaul. One of their most attractive features is that for a normal fast road car they can, by changing the jet sizes, be set to suit an engine from standard power output up to a level which most people find sufficient for normal fast-road tuning. Overhauling the popular Weber DGAV 32/36 is covered in a later chapter.

◀ *If your tuned Crossflow is going in a kit car with a remote header tank, one modification should be the thermostat housing from a Mk I XR2. It won't help the power but it will give you a handy filling point.*

For competition power you move into the more specialist realm of racing and would have to go for a pair of carburettors, possibly Weber DCOE 40s or 45s, on special manifolds.

Even if you don't have any porting done on the head, you'll need a better exhaust system. For anything from mild up to medium-fast tuning there are good proprietary systems you can buy from people like Janspeed or Peco but, for higher power, you need to go to a specialist exhaust builder.

All these points are covered in detail later on in this book.

Workshop and tools

Now we come to something which is just as important as what you do to the engine: where you do it and the tools you need to do it. Stripping, cleaning, tuning and rebuilding an engine needs room. Parts of the job, particularly the stripping down and cleaning, can be messy. A small garden shed which you have to share with garden tools, the mower, a garden hammock and the children's bikes isn't ideal. You want somewhere where, at the end of a session, you can put parts down, clean up and leave everything undisturbed.

To most people this means taking over at least half the garage, where your minimum requirements will be electric power, a good sturdy bench and a decent-sized vice.

When it comes to tools, I am not going to make out neat lists because no-one of my acquaintance ever goes out and buys all the tools listed in books on home workshops. Having said that, there are minimum requirements. It can be frustrating, to say the least, if you have to stop work to go and buy a needed tool – so try to plan ahead as much as possible.

To take the engine out of the car, and replace it when it's finished, you will need either an engine crane, or a hoist, and either a shear leg frame or a strong beam across the roof of the garage. Remember that when you are actually lifting the engine the load on the beam is twice the dead weight of the engine. Make sure that you can lift the engine high enough to be able to push the car out from under it.

For washing down the outside of the engine before you start stripping it you'll want a tray big enough to stand it in and, for cleaning off the parts when they're dismantled, you'll need a good-sized tank or tub big enough to stand the block in. A rack or shelves, or an old table, to lay out the parts when they're cleaned off sounds like a luxury and I've known people rebuild engines with just a collection of cardboard boxes to store the parts, but it makes the job much easier and more pleasant if you have somewhere to lay them out. You will also need a good supply of old rags for wiping down.

Depending on how much of the work you do yourself and how much you put out, it's possible to rebuild and tune an engine with a surprisingly small number of tools. Most people who undertake rebuilding an engine will already have a basic tool kit of socket set, spanners, screwdrivers, pliers and so on, and I will mention some of the other, less frequently used, tools you'll need as we come to them. Here I would stress that you should get the best tools you can afford. Very cheap spanners and sockets are made from inferior steel and don't last. They wear quickly and slip, resulting in rounded nuts and bolt heads, skinned knuckles and inferior work. Good quality tools are a pleasure to use and make the job so much easier.

As well as a good quality socket set you will need a torque wrench. Torque wrenches come in various sizes, and in workshop manuals you will find some torque tightening figures as low as 8 or 9 lb ft. Ideally you should have two torque wrenches to cover the various torques quoted. If you can afford only one, make sure it covers the main and big-end bearing caps, the flywheel and the cylinder head.

Most professional mechanics have two torque wrenches, one for high and one for low torques, but you can get away with one on a Crossflow. The highest important torque you will find is the final tightening of the cylinder head bolts to 71 lb ft. You need a wrench which is not working at its top limit, so I suggest you get one which goes up to 100 or 120 lb ft. At its lower setting it should come down to about 20 or

25 lb ft, or maybe lower depending on the make.

The type which 'breaks' with a click when it gets to the set torque is much easier and more positive to use than the type which flexes and has a pointer moving across a scale. They are accurate enough, but you have to be careful in reading them instead of just setting the torque and tightening till the wrench clicks.

As an alternative to doing all the work yourself you can go along the path used by many home engine tuners and have part of the work done outside. Most tuning firms are happy to recondition your engine, or supply an exchange, to what is generally termed a 'short' engine. This is a complete rebuild of the crankcase, including reboring, crankshaft grinding, and so on, to whatever stage of tune you want. Similarly, they offer exchange cylinder heads tuned to what is generally termed Stage 1, Stage 2 or Stage 3.

Different firms have different definitions of exactly what these various stages are. Some don't use the terms at all, preferring to give their tuned heads names such as fast-road, full-rally, and so on, so it's little use going to a tuning firm and saying, 'I want a Stage 2 head.' If you want to buy a completely tuned head ready to bolt on, get leaflets and brochures from various tuning firms and compare the specifications, as well as the names and prices. However, many people still talk in terms of Stage tuning, so I'll run briefly through the more usual interpretations.

Generally speaking, Stage 1 is designed to give a moderate increase in power to use with a free-flow air filter and a performance exhaust. The increase in power will make the car feel much more lively and responsive, but it isn't usually enough to warrant uprating the bottom end, though most tuning firms advise at least a change of bearing shells and an uprated oil pump. If the bores and pistons are not badly worn, you can sometimes get away without a rebore, though new rings are advisable, as will be a complete overhaul of the rocker gear.

Stage 2 takes the tuning higher, usually with increased diameter inlet valves, and is designed to go with a performance camshaft and vernier timing gear. To get the best from this stage of tune you need to recondition the engine with a rebore, new pistons and crankshaft regrind, plus new rockers and shaft and a higher capacity, higher pressure oil pump. Most tuning firms recommend having the crankshaft and flywheel balanced, and changing over to higher duty bearing shells.

Stage 3 is sometimes called full-rally, or even race tune. A head like this would be completely wasted on an engine that wasn't rebuilt to take the higher power. This is where you move into the realms of checking main bearing bores for ovality, and machining if necessary, checking the crankshaft grinding for equal throw, checking connecting rods for equal throw and ovality of the big ends, balancing connecting rods, balancing pistons, a pair of Weber DCOE carburettors, and so on. The result isn't generally an engine that's comfortable being driven round town and doubling as a family car, and it isn't a stage of tune to take on at home unless you are an experienced engine builder, or unless you have the help and advice of a tuning firm.

I will talk about these higher stages of tune later in the book, but we are not aiming at exotica. In the main, we are aiming at home rebuilding to give you an engine that is much more lively, much more responsive and much more pleasure to drive than a Crossflow in standard factory tune.

Chapter 3

Getting more power: the general principles of tuning

Tuning for power is a complex subject and you can, if you want, read very learned papers about combustion inside the cylinder head. Unless you have degrees in mathematics and physics you may not be much the wiser but, fortunately, when it comes to practice, a lot of it can be simplified.

Designers of standard production saloon engines have to compromise because they are faced with conflicting requirements. They've got to provide enough power to stop such cars being sluggards and they've also got to please Mr and Mrs Average Motorist who want that power delivered as smoothly and quietly as possible. They also want reasonable power over a wide rev range so that they don't have to row the car along on the gear lever to get decent acceleration.

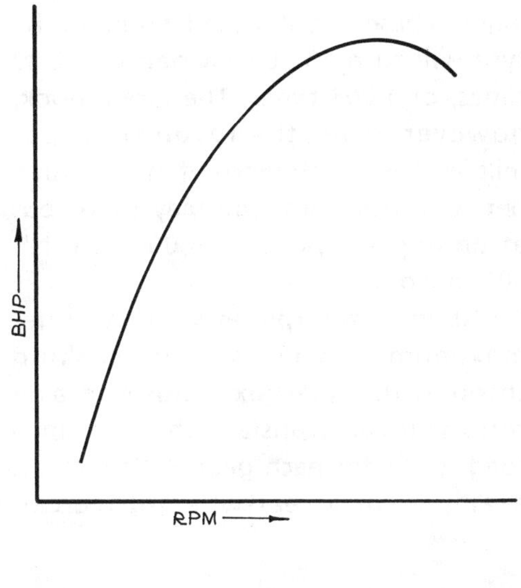

▶ *A typical power curve for an engine designed for easy motoring without much gear changing. Top power isn't very high, but the curve is gentle to give quite a lot of power lower down the speed range.*

This means that the designer has got to provide a power curve that doesn't peak too sharply, so that all the power is at the top end of the rev band, and the torque curve must be reasonably flat.

The simplest way of describing torque is the turning force on the crankshaft, and power and torque cannot be separated because power depends on torque. The basic equation for horsepower is:

$$\text{Horsepower} = \frac{2 \times \pi \times N \times T}{33{,}000}$$

where N is the rpm of the engine and T is the torque. The figure 2, the mathematical constant π and the figure 33,000 are all constant, so that power depends on the torque multiplied by the rpm. We can get more power by increasing either or both.

In practice, we have to be careful whereabouts in the rev range we increase the torque. Look at the power and torque curves of two widely different engines. The first (see left) is a real slogger, plenty of torque at low revs and a fairly flat torque curve. The power curve isn't particularly high but it's a nice gentle curve with quite a lot of power at medium revs. They are the sort of power and torque curves you had with the big, lazy side-valve American engines of a few years ago. With this sort of engine you could put the car into top gear at low speed and leave it there practically all day. The acceleration wouldn't be anything to shout about, but it could pull a fairly high back axle ratio, three ratios in the gear box were ample and it did its job smoothly and without any fuss.

Now look at the power and torque curves of a very different engine (see right). Here, the designer has gone for out-and-out power. The torque and power at low revs are poor and the tickover at anything below about 3,000 rpm would be lumpy and uneven. If you tried to put a car with this engine into top gear at low speed it would kick and buck and be almost undriveable. But it's a very different story at high revs. The torque increases and the power curve shoots up. It's a real screamer of an engine which will give the car a high top speed and searing acceleration, provided the driver has enough gear ratios to keep the engine at high revs over a wide road speed range. Fine for sprints or circuit racing, but it would be hard work to take through town traffic.

At higher power where you tend to get a peaky power curve, moving the maximum power along the rpm axis may mean that you need to consider changing the rear axle ratio or the gearbox ratios, or fitting a five-speed gearbox. You can see the effects of this by drawing a new graph, derived from the power curve but substituting road speed along the horizontal axis in place of engine rpm.

Before you can draw this graph you will have to decide what tyres you are going to fit to the drive axle. Changing wheels and tyres is a very popular pastime on tuned Fords and before you can start calculating engine speed against road speed you need to know the wheel turns per mile of the tyres you are fitting.

You can't work this out accurately from the normal outside diameter of the tyre nor from the static laden radius

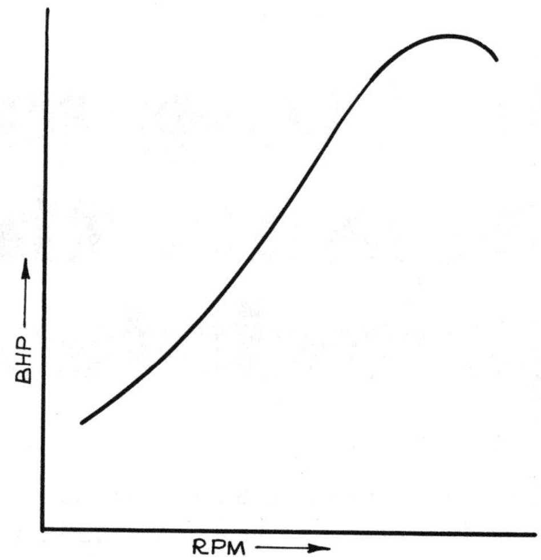

◀ *In contrast with the first power curve, the curve on this highly tuned engine is steep and peaky. The power falls off very sharply in the lower speed range but, at high revs, it shoots up. A real screamer of an engine designed for racing.*

because tyres grow in diameter at speed. You must get the revolutions per mile as quoted by the tyre maker. The place where you buy your tyres should have the maker's data book which gives this.

As an example of how far out you can be if you try to calculate it, let's look at a popular upgraded tyre on an Escort – a 185/60 R15, which is a low-profile replacement for the standard 155R13. I'm using a Pirelli book here, but other makers issue similar books, though the figures may vary slightly from maker to maker. The Pirelli P600 185/60 R13 has a static laden radius of 254mm, or 10 in, which is an equivalent diameter of 20 in and an equivalent circumference of 62.83 in. There are 63,360 inches in a mile, so based on static laden radius, the tyre will turn 63,360 divided by 62.83 times, or 1,008 times. The Pirelli book, however, gives the revolutions per mile as 946, a difference of more than 5 per cent. Not a lot, you may think, but at an engine speed of 6,000 rpm it's 300 rpm out.

You know the rpm at which you get maximum power, so by working through the gearbox ratios and axle ratio you can translate this rpm into road speed for each gear. Let's take an example for a rear-wheel-drive car.

▶ *If you change the base axis of the graph to road speed instead of rpm, you can plot a series of curves to show where the power occurs in each gear. This will give you an idea of the best speed at which to change gear when you're accelerating to avoid a 'surge, drop' type of acceleration.*

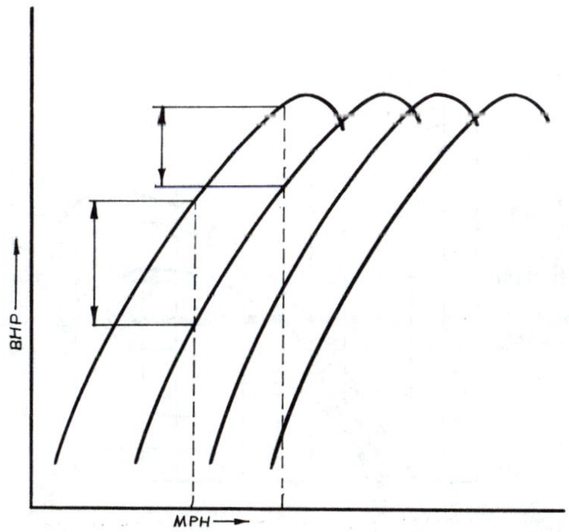

Suppose you get your maximum power at 6,000 rpm, then, with a 1 to 1 top gear ratio, the propshaft will also be doing 6,000 rpm. If the drive axle ratio is, for example, 3.54, the drive shafts, and thus the wheels, will be doing 1,695 rpm. A tyre with a revolutions-per-mile figure of 946 will travel 1,695 divided by 946 miles in a minute, or 1.792 miles per minute, a theoretical top speed of 107.5 mph.

▶ *A change to a performance camshaft can really boost the power, as this graph from Piper RS shows. The top curve at about 105 bhp is Piper's Blueprint 285 cam, the second curve is the BP 270 at around 95 bhp and the lowest curve at 84 bhp is the output for a 1600 Sport engine with a standard camshaft. Note that the performance camshafts give virtually no loss of power in the lower rpm range. The engine was tested with the carburettor in standard tune.*

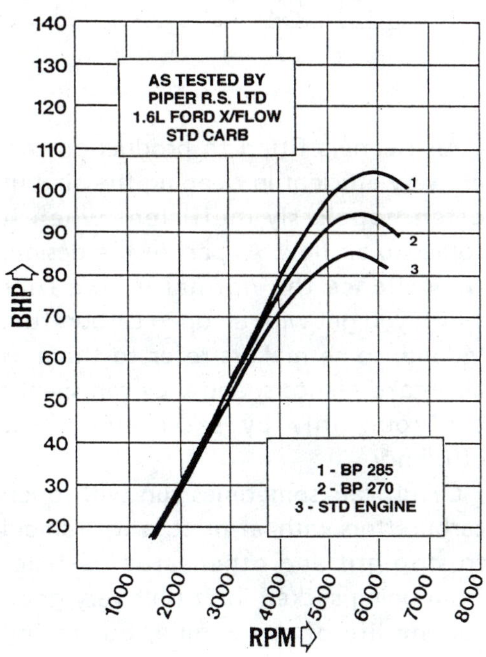

Whether or not the car will be able to achieve this road speed depends on how much power there is available at 6,000 rpm to overcome rolling resistance and wind resistance at over 100 mph. Wind resistance is something you can't calculate without testing in a wind tunnel, but it increases in the ratio of the square of the road speed so the last 10 mph to top speed takes a lot more power than a speed increase of 10 mph at a lower speed. You may even find that your car is faster in third gear than in top. Almost certainly it will be faster in direct top gear than in overdrive fifth if you fit a five-speed box, but it will be a lot less fussy and more economical at 70 mph with a five-speed box.

To return to our graphs, the shape of the power curve will be the same in each gear so you just repeat it along the mph axis of the graph spaced according to the gear ratios to get a peak power mph figure.

This is a very useful curve for sprints and racing because you can draw a pair of horizontal lines, as I've shown in the diagram above, which give you the best road speed range at which to change gear to avoid the surge-drop type of acceleration you get during acceleration through the gears if you change gear at the wrong point. If you prefer to work from the rev counter rather than the speedometer, put these values back in the power vs RPM curve to give the best range of engine revs in which to change gear.

For road use you can use these two graphs to predict the behaviour of the car on the road. Suppose, for example, that you are driving along a normal main road (not dual carriageway) where the speed limit is 60 mph and you catch up with someone doing between 45 and 50 mph. You have to slow down behind them to wait for a gap in the oncoming traffic and then you want to overtake and get back on the left side of the road as quickly as you can. Is it best to leave the car in top gear, change down a gear, or change down two gears?

Looking at the power vs road speed graph will tell you what power you have available at any speed in any gear and whether or not it would be faster to change up again during overtaking or leave it till you have completed the manoeuvre.

Changing the rear axle ratio or the tyre size will move the whole group of power curves bodily back and forth along the road speed axis. Changing the gear ratios inside the box, or changing to a five-speed gearbox, will alter the spacing between the power peaks in each gear, so you can predict what effect such a change will have on the car's performance and whether it is worth doing before you spend a lot of money on parts to find out on the road.

Basically, the only thing that gives you power is the fuel and, to get the power from it we need to vaporize it, mix it with the correct amount of air, get this mixture into the combustion chamber and ignite it. It follows that the more air/fuel mixture we can cram into the combustion chamber, the more power we get.

Turbocharging or supercharging is one way of doing this, by compressing the mixture on its way to the cylinders, but there's a great deal we can do with a normally aspirated engine where the mixture is sucked into the combustion chambers when the piston goes down on its intake stroke. Most tuners refer to this as improving the breathing of the engine.

Anything which slows down the flow of mixture into the cylinders is a bad thing when we're looking to increase the power, which is why in the first stages of tuning we concentrate on the intake side of the engine.

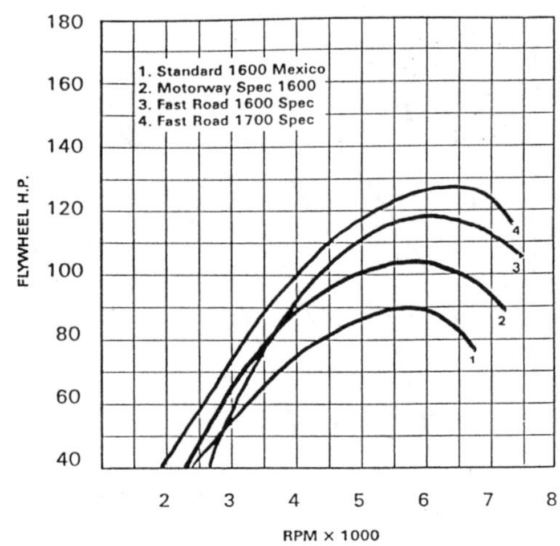

◀ This shows the output curve of the standard Escort 1600 Mexico compared with three versions of the engine tuned by Oselli Engineering. As expected, the Motorway and the 1600 Fast Road tunings both boost the top power, though the higher output of the Fast Road specification means that it loses slightly at rpm below 3,500. But look what a difference an increase in capacity makes. The Fast Road 1700 (0.090 in overbore) engine comfortably tops the rest all the way up.

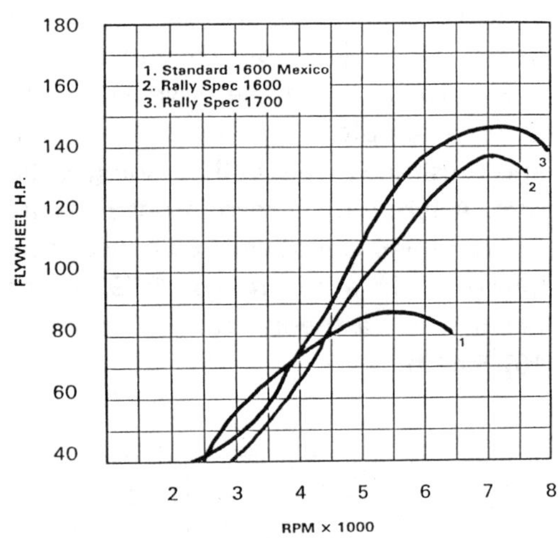

◀ The curves take on a completely different shape when you start pushing the power higher. These are also Oselli tuned engines with the 1700 Rally Specification nudging 147 bhp. The curve falls off and becomes concave below 4,500 rpm and more or less mirrors the lower capacity 1600 Rally engine. Neither engine, however, has the low-down power of the standard 1600 Mexico engine below 3,500 rpm. If you want it at the top you can't have it at the bottom as well.

Air cleaners fitted to production cars are very efficient in cleaning the air, but often hopelessly inefficient when it comes to air flow. As part of the design, they silence the normal intake roar which you get with an open carburettor. Indeed, some makers refer to them as air intake silencers, and they can silence the roar only by sacrificing flow efficiency.

Circuit cars sometimes run with open carburettors with, at most, a wire shield to stop grit and other large particles from being sucked in. It isn't very good for the life of the engine, but racing engines are stripped down for rebuilding far more frequently than the engines of road-going cars. For a fast road car we have to strike a compromise, which is why companies like K&N, Jetex and Megaflow produce what are generally termed free flow filters.

They aren't completely unrestrictive, but their flow figures in cubic feet per minute, or cubic centimetres per minute, are much higher than a normal production air filter/silencer. They dampen less of the intake roar, but that's one of the penalties we have to put up with. Some people find that, in winter, a free-flow air filter tends to become blocked with ice but, as we will see later, there are ways of overcoming this.

Changing carburettors is a favourite with home tuners. Ford has used three types of carburettor on its Crossflow: Motorcraft, VV (Variable Venturi), both made by Ford, or one from the Weber range.

The Motorcraft is a good workhorse type of carburettor and seldom gives any trouble, but for higher power it has got its limitations. Most tuners won't touch it. Ford's VV has a bad, somewhat unwarranted, reputation for giving trouble. It's a lovely carburettor when it's on song, and you would be hard pressed to better it for a combination of power and economy. Unfortunately, it is complicated, and when it gets out of adjustment it can sometimes be a real pig to tune. Those tuners who have come to understand its foibles, like it, but for home tuning I would recommend switching to a Weber.

On small Crossflows, 1100s and some 1300s, Ford used a single-choke Weber, which really comes under the heading of 'cooking carburettor'. Like the Motorcraft, it's a good workaday instrument but it isn't very tuneable.

Much to be preferred is one of Weber's double-choke (sometimes called twin-choke or twin-barrel) carburettors – with one exception. Many home tuners use a Weber double-choke carburettor from one of Ford's other engines, usually from the overhead cam Pinto, because these can be picked up at a reasonable price and are quite easy to rebuild at home if they are not too worn. The exception, which I wouldn't recommend, is the one from the Sierra with the E-Max 1600 Economy Pinto engine. Unlike most other Weber double-choke carburettors, this has a vacuum-operated second barrel and is much more difficult to set up as a power carburettor than those with a mechanically-operated second barrel. I deal with carburettor swapping and rebuilding later in the book.

Next in line on the intake side is the induction manifold. Here, standard production designs vary from quite good to horrible. The standard manifold on the 1600 GT or Sport Crossflow engine isn't at all bad and, unless you are going for top power – in which case you'll probably be fitting Weber DCOE horizontal carburettors with special manifolds – I would advise staying with it. Equally good for 1300 engines (and even 1100 engines) is the manifold on the old 1300 GT Escort Mk I Crossflow, even though on the earliest versions the engine block itself wasn't as robustly

▼ *If your Crossflow has a Motorcraft or VV carburettor you will want to make a change when tuning. One of the most popular changes for fast road work is to a twin-choke Weber DGAV.*

stiff as the later engines.

Alternatively, you can opt for a proprietary manifold. This will give you better air flow figures than Ford's and, if you have money to spare, then by all means get one. If, on the other hand, money is tight, you may be better off sticking with a Ford GT or Sport manifold and spending the saving elsewhere on the engine.

Quite important on the inlet manifold is blending the outlet ports of the manifold to match the intake ports on the cylinder head. Recent research on flow benches has shown that it isn't quite so important as was once thought, but it's still a good general principle of gas-flow tuning that the inlet ports on the cylinder head should be slightly larger than the outlet ports on the manifold to avoid a ridge that might set up swirling eddies that can play havoc with the gas flow.

That brings us to porting and gas-flowing the cylinder head itself. Ford's Crossflow cylinder head is very amenable to tuning. In old books on head porting you might read about people polishing the ports on a cylinder head to a smooth mirror finish. That used to be accepted theory, and it's an accepted fact in physics that a smooth bore offers less resistance to gas flow than a rough one.

However, we've learned a lot about intake gas flow since those days. Most tuners today leave the walls of the ports with a fine matt ground finish. The thinking behind this is that, however good it is, a carburettor only atomizes the fuel into very fine droplets, rather than actually vaporizing it, so strictly speaking our 'gas' isn't a gas at all, it's air with very fine fuel droplets suspended in it. With a mirror-finish bore, these droplets tend to cling to the sides till they drop back into the air stream as bigger droplets, which upsets the mixture. With a matt finish bore the droplets are much less inclined to drop out, and any that do bounce back into the air stream so that the mixture remains more constant.

Far more important than bore finish is the shape of the port and, once again, I'll deal with that later. If money is very tight, you can do a lot to improve the gas flow of your cylinder head and manifolds at home, but for anything more than a mild pep-up entrust it to one of the specialists.

Many books on tuning deal at length with swirl inside the combustion chamber, and compression ratio. There's no denying their importance but, with Crossflow engines, you haven't got a combustion chamber in the cylinder head – at least not in most of them – it's in the piston crown. You have to rely on the makers of specialist pistons if you want to increase the compression ratio, or go to one of the specialist tuners who machine pistons.

This is essential if you are going for larger diameter valves as part of gas-flowing the head. With larger valves, particularly if you are also fitting a performance camshaft with higher lift, there is a distinct danger of the valves hitting the tops of the pistons with disastrous results.

Performance camshafts, from people like Kent Cams or Piper RS, can give your engine a real boost in power. Choose carefully, though, as there is a lot more to a performance camshaft than just increased lift. Don't go by word of mouth or advice from a friend of a friend. Talk to the camshaft maker and to the specialist tuner if you are having your head gas-flowed. Discuss with them the sort of power boost you are looking for and the camshaft you will need to achieve this. Later I'll go on to changing rocker ratios, which can also give increased lift and, for top power, roller bearing rockers.

The vast majority of Ford's Crossflow engines were made before the advent of unleaded petrol – perhaps 'second coming' would be more appropriate than 'advent' because all petrol was unleaded 70 or 80 years ago – and they were designed to run on leaded fuel. Some low compression versions were designed to run on two-star or three-star fuel, but the majority should use four-star.

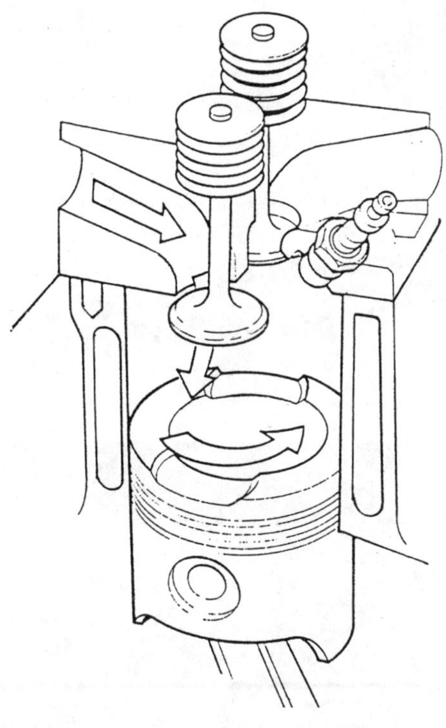

▶ *Swirl of the incoming mixture is an important part of good combustion. Because the Heron bowl-in-piston chamber can't give swirl on its own, the swirl has to be given by the angle and shape of the inlet port.*

If you are having the head gas-flowed, it is worthwhile at the same time having it converted to run on unleaded fuel. Quite apart from saving on running costs, there will come a time when leaded fuel becomes more difficult to obtain.

If you don't have the head converted, there's a big temptation to try unleaded. At the time of writing there's a price difference of 13p to 14p a gallon, depending on where you buy it, between leaded and unleaded. When you look at the octane ratings you find that premium unleaded is the same or slightly above the rating for leaded fuel.

There's a lot of misinformation about which engines can and can't use unleaded fuel. Some people say you can if you retard the ignition. Others say you can if you use a mixture of one in four leaded to unleaded and, if you look at charts in garages, Ford seems to agree with this.

There are various gadgets and things you can hang in the petrol tank which, their makers sometimes say, allow you to run on unleaded all the time. Then, of course, engine tuners and specialists will tell you that the only safe way to use unleaded in an engine that was designed to run on leaded fuel is to have the head modified. So who's right?

Lead compounds were first added to fuel back in the 1920s as an anti-knock agent when compression ratios started to go up as designers looked for more power. In those days it was common practice to have the head off every so often, usually around 10,000 to 15,000 miles or so, for a decoke and valve grind and, even though engines were very lightly stressed and valve springs were quite weak compared with today, pocketed valve seats and worn valve heads were quite common.

When lead was added to the fuel it was found that valves and valve seats lasted much longer because, when they burnt, the lead compounds left a deposit which acted as a lubricant. Over the years, powers and compression ratios went up and, though the steels used for valves improved, they didn't change all that much. Even so, valves seating directly in plain cast-iron heads lasted a long time thanks to the lead.

There the situation stayed until it was realized how much harm burnt lead compounds in the air could do to us. Other compounds that didn't contain lead were introduced as anti-knock

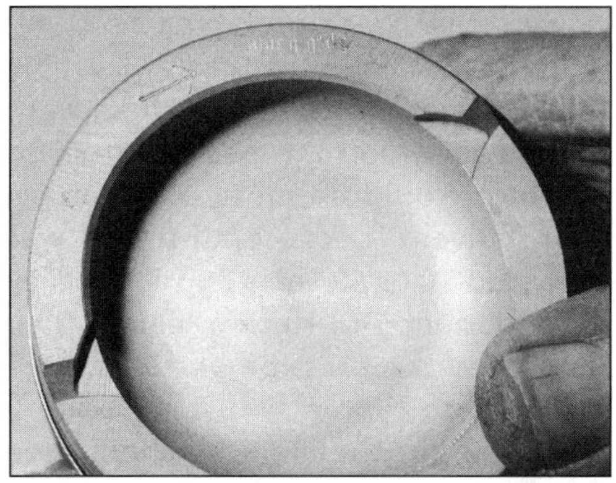

◀ *On a flat-head Crossflow the combustion chamber is wholly in the crown of the piston, so compression ratios can be altered by changing the pistons.*

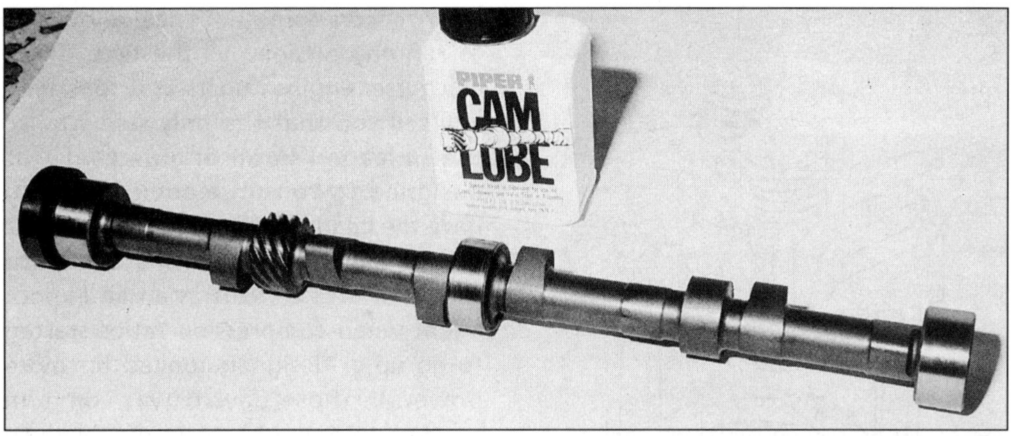

◀ *A change of camshaft to give greater lift is probably the biggest single change that you make to boost the power of your engine in a home garage without recourse to outside gas-flowing of the head or over-boring.*

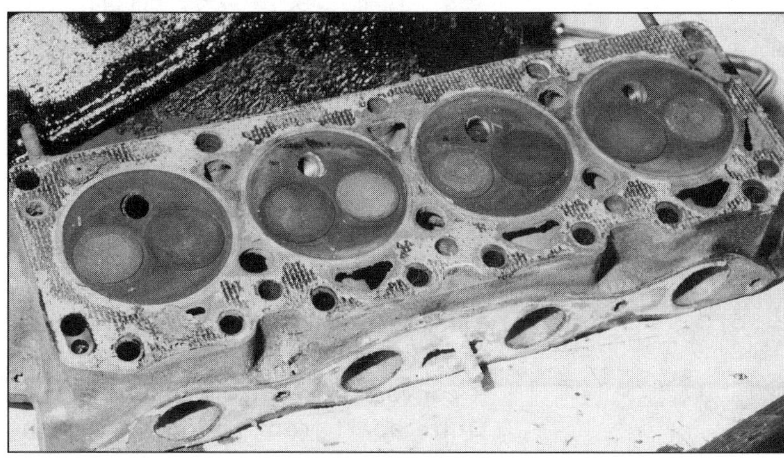

▲ *Modern fuels with detergents do help to keep the inside of the engine clean. The carbon deposits on this head are light and even and quite soft. With older fuels we would have expected an engine at this mileage to have clumps of hard carbon inside.*

◀ *The only safe way to change to unleaded fuel is to use valves of a better grade steel, and have hardened exhaust valve seat inserts. The head has to be lightly faced off afterwards as the inserts stand proud when they are put in.*

agents and we had unleaded fuel. Because these new additives didn't lubricate and protect the valves and seats like lead, engine designers had to push up the quality of steel used for valves and put in hardened steel seat inserts.

If you use unleaded fuel, whatever the octane rating, in an engine designed to run on leaded fuel you will damage the valves and valve seats.

So, what about this business of retarding the ignition? If the engine already has valves and valve seats suitable for unleaded, but four-star leaded or super-unleaded was specified because of its anti-knock properties, retarding the ignition will in some cases allow you to run on the cheapest 'ordinary' unleaded.

What about the one-in-four ratio which Ford say you can use? Ford carried out a lot of research into this and found that if their valves and valve seats had a dose of lead compounds for a quarter of the time, they were quite happy. But, and this is the big but, that one-in-four dose had to be undiluted leaded fuel.

If you read Ford's recommendation carefully, you'll see that they don't say you can mix leaded and unleaded one in four, they say one tankful of leaded in every four. The amount of lead in leaded fuel has been reduced a lot in recent years and, if you mix it with unleaded, the amount of lead your valves get just isn't enough to protect them. In other words, you've got to run the car practically out of fuel before you fill up with leaded, run it very low and then keep a careful note of how much unleaded you put in and run it practically out of fuel again after running on three tanks full. Not very convenient is it?

Ford, by the way, was talking about standard production engines, not tuned engines.

What about gadgets and additives? There's been a lot of controversy about this, even legal action, so I am not going to add to it except to say that some people who have carried out tests say that they don't give your engine the equivalent of the lead protection the valves and seats need. I wouldn't want to chance it while there's any doubt because it takes a lot of 14p a gallon saving to buy a new cylinder head.

So, then, what's the answer? My advice is that if you've got an engine that was designed to run on leaded fuel, stay with leaded fuel unless you have the head converted. All the well-known engine tuners will convert your cylinder head with valve seat inserts and higher quality steel valves and guides. Some offer an exchange over-the-counter service.

If you're in any doubt, don't rely on the year of registration or manufacture of your car. Ford introduced unleaded engines over a period of time on the production line and identified them by different engine codes. Your local Ford main dealer will have a list of these codes and he will be able to tell you if it's safe and whether or not you have to alter the ignition timing.

Discuss this with the person who is gas-flowing the head. The very minimum you will need to have done is to have the exhaust valve seat machined out, a hardened valve seat insert put in and better quality steel valves. You will also have to discuss with him, possibly together with the piston manufacturer, if you are going for increased compression ratio, how far up you can push the compression ratio before you start getting pre-ignition, pinking or detonation, any or all of which can ruin your engine. This is something of a grey area with many people because it's also tied up with ignition timing, and all three tend to get lumped together as 'engine knock'.

If you delve into books about engines from the vintage era you sometimes see the phases of the four-stroke cycle described as 'induction, compression, explosion, exhaust'. To irreverent mechanics this soon became 'suck, squash, bang, blow'.

Engine designers and fuel chemists didn't like this at all. Not because they were against the use of slang or jargon, but because it implied that the fuel mixture burned with an explosive bang, and this was one of the things they were all trying hard to avoid.

The fuel mixture in your combustion chambers ought to ignite and burn in a controlled way. If it doesn't – in other words, if you get an explosion, or bang – you're getting detonation or pre-ignition, either of which can do your engine a great deal of no good.

Both detonation and pre-ignition produce similar sounds – a high-pitched knocking – and they're often confused. In fact, they're quite different, and even though you sometimes get pre-ignition brought on by detonation, they're brought about by different circumstances.

Pinking – another term sometimes used to describe combustion knock – isn't really the same as either detonation or pre-ignition. Pinking, by itself, is brought on by having the ignition too far advanced. The mixture fires and almost completes its burning before the piston gets to top dead centre. The piston tries to compress this rapidly expanding mass of gas and the shock loading on it is very high. Sometimes, at low revs it gives up and kicks backwards but, at speeds higher than a slow tickover, the flywheel carries it over top dead centre and the sudden sharp extra load sends a shock wave right through the engine which you can hear as a high-pitched metallic-like knocking.

Pinking is always cured by retarding the ignition. Detonation is very close to pinking – so close, in fact, that I prefer to lump the two together because it's very difficult to say when one changes to the other. Pre-ignition is something different, as I explain later.

Let's look first at detonation. This happens when part of the mixture suddenly burns with a bang – in effect, a spontaneous explosion – after the fuel has been ignited by the sparking plug but before it has all burned in a controlled way.

What happens is that when you get ignition, the compressed mixture in the combustion chamber doesn't all burn at once. It's all over in a few milliseconds but, to start the burning, the sparking plug ignites the fuel mixture immediately around it and, as the mixture burns, you get a flame front moving across the rest of the mixture.

This flame front is the front wall of a mass of rapidly expanding and very hot burning gas and, as it moves across the combustion chamber, it compresses the unburnt gas in front of it. This unburnt gas doesn't like being compressed even more, and it will (given the chance) ignite all by itself before the flame front reaches it. It doesn't burn in a controlled and civilized way, but goes off with a bang, and that bang is the knock you hear when you get detonation.

OK, we know what it is, so what causes it and how can we avoid it? Most people know two parts of the answer – one being to retard the ignition and the other to use a fuel with a higher anti-knock rating. Ignition timing is, naturally, tied up with engine speed. After the plug sparks, the flame front travels across the combustion chamber at a constant speed and the ideal time to complete the combustion process is when the piston is just leaving top dead centre on its way down for the power

Detonation and local overheating can play havoc with sparking plugs, as these pictures from Champion test engines show. On the plug on the left, the centre insulator has cracked and, on the one on the right, both electrodes have burnt away.

stroke. To do this, we have to start the ignition going before the piston reaches TDC and, as the engine speed increases, we have to spark it earlier and earlier, which is why we have automatic advance on a distributor.

If we don't advance the ignition, the burning happens later in the engine's cycle and there is less chance of detonation, but the piston has already started on its downward stroke before it gets the full benefit of the expanding gases so we lose power. This critical timing of the ignition to get the full benefit of the energy in the fuel we burn without going to the point where the early spark increases the risk of detonation is a too often under-rated part of engine tuning.

The trick is to get it right over the whole speed range of the engine. Even then, it isn't just a simple case of matching the advance curve to the speed. The power required at different speeds also has a bearing on it because the more power we need the more fuel and air mixture we require.

If, at any part of the speed range, we ignite the mixture too early, the mixture is still being compressed by the piston as the flame front moves across. This adds to the tendency of the unburnt gases to ignite spontaneously and increases the chance of detonation.

In most cases, the detonation we're all familiar with happens when we put the engine under heavy load with plenty of throttle opening at low to medium rpm – for example, when accelerating uphill. It can certainly damage the engine, but not so much as high speed detonation at maximum torque and rpm above that. Detonation here doesn't make itself so obvious as detonation at lower speeds, but it can do far more damage to the engine.

Not only does it set up shock waves in the engine, which stress it mechanically, there is also the heat factor. When detonation happens, the local temperature in the area where it occurs rises steeply. Usually this is the area in the combustion chamber round the exhaust valve. If you strip down an engine which has been suffering from high speed detonation you often find burnt exhaust valves and burnt pitting marks round the edge of the piston, and pitting of the head round the exhaust valve seat.

This sudden rise in temperature and the harsh explosive nature of detonation also affects other areas of the combustion chamber, in particular the sparking plug. At certain high speeds, the effect of the high frequency detonation shock waves can cause the earth electrode of the plug to vibrate like a tuning fork. If it goes on long enough, metal fatigue sets in and the electrode can suddenly break off. The plug also becomes overheated, the electrodes burn away and, more often than you think, the insulation round the centre electrode can crack and flake off.

There's yet another effect of local overheating – pre-ignition. This is ignition of the compressed mixture before the plug sparks. It has the effect of suddenly advancing the ignition, which increases the detonation so we can get a vicious spiral effect.

The pre-ignition is caused by something inside the combustion chamber reaching a temperature where it glows and starts the compressed mixture burning. It could be a plug electrode, it could be a thin edge of an exhaust valve or a piston cut-out, or it could be lumps of incandescent carbon.

Fitting a camshaft with higher exhaust valve lift and a longer duration off its seat increases the danger of the exhaust valve overheating. Unlike the inlet valve, which is firmly on its seat during the hottest part of the combustion cycle and has cold incoming gases on the underside of its head to help keep it cool, the exhaust valve head gets very little help in the way of keeping cool.

It's pushed out into the combustion chamber during a very hot part of the combustion cycle and the only chance it gets of getting rid of its heat is the very short time it's in contact with its seat. In the days of high-performance piston-engined aircraft, engines like the Merlin had exhaust valves with hollow stems. The stems were filled with sodium which became liquid at high temperatures and, as it splashed up and down the inside of the stem, helped to conduct the heat away from the valve head.

We don't use sodium cooled valves on car engines but, on many tuned engines, we use exhaust valves of a higher grade of steel alloy and, unlike an inlet valve where a thin band of seating is quite adequate, the seat ring on an exhaust valve should be reasonably wide.

With regard to plugs, you should always follow the plug maker's advice about the heat range of the plug you use. The type specified by Ford for normal, untuned engines is quite adequate for its job, but put that plug into the much hotter conditions of a performance tuned engine and it can't get rid of its heat quickly enough. You need a 'colder' plug, one in which the hot centre electrode can get rid of its heat to the plug body more quickly.

You can recognize 'cold' and 'hot' plugs by looking in the end. If the ceramic insulation round the centre electrode is in contact with the body almost as soon as it clears the tip of the electrode, and there's hardly any space inside, it's a 'cold'-running plug, designed to be used in engines that reach high internal temperatures. If the ceramic insulation tapers off gradually and joins the body much higher up leaving plenty of room inside, it's a 'hot'-running plug designed for engines where the internal temperature is lower.

So why don't we all fit very cold plugs and be done with it? It isn't just a question of price, though cold plugs for high-performance engines are often more expensive because fewer of them are sold. A cold plug doesn't perform so well as a hot one at low engine speeds. It tends to foul up more and you get slow running misfiring. This, as well as wild cam profiles that don't like slow running, is one of the reasons why racing drivers blip their engines on the starting grid.

There's another factor which affects the chance of detonation: proper mixing, or otherwise, of the fuel and air. In this respect, though not in others, carburettors are often superior to fuel injection because the mixing takes place earlier and the air and fuel has a chance to complete its mixing in the inlet manifold, often in a plenum chamber just under the carburettor – one of the reasons, as well as gas flow, for fitting a good design of inlet manifold on a performance engine.

Air and fuel which isn't mixed properly has a much greater tendency to detonate than if it is thoroughly mixed. Combustion chamber design also has an important part to play. Mixture, which just rushes into a combustion chamber and suddenly stops, is static when ignition takes place and there is much more likelihood of detonation than with a mixture which keeps on moving, or swirling, until it burns.

The Heron, or bowl-in-piston, combustion chamber on the Crossflow doesn't in itself have very good swirl characteristics. All the swirl has to come from the angle and shape of the inlet port. Unfortunately, well-meaning but ill-informed gas-flowing of the inlet port can upset the swirling movement of the mixture inside the combustion chamber. Yes, gas flow figures are important in the search for more power, but unless they're combined with good swirl you get a risk of detonation which means that you can't use the power without risking damage to the engine. The better head tuners may charge more than others, but you're buying experience in all the aspects of combustion, not just gas flow.

Before I move on to fuel and octane ratings, we also have to consider compression ratio. It should be obvious that the higher the compression ratio the higher the pressure inside the combustion chamber at the time of ignition. Broadly speaking, the higher the pressure, the more power we can extract from the fuel, but the more pressure the flame front exerts on the unburnt gases so the greater the chance of detonation. It's a vicious circle, isn't it?

So far I've talked only about the engine and its effect on detonation, but fuel, of course, plays a vital part. The anti-knock, or anti-detonation, quality of a fuel is expressed as its octane rating. The higher the number, the less likely it is to detonate. Octane numbers are measured in fuel laboratories by comparing the anti-knock properties of the fuel against a mixture of iso-octane and n-heptane. These chemical names may not mean a lot to us, but the more iso-octane in the mixture the better it resists detonation. The proportion of iso-octane in the mixture, expressed as a percentage, that gives the same resistance to detonation as the fuel under test is the octane number.

You often find two octane numbers quoted for the same fuel. One is the Research Octane Number, usually abbreviated to RON, and the other is the Motor Octane Number or MON. The number you see quoted at fuel pumps is nearly always the RON number and it's measured under reasonably mild test conditions such as you might find in an engine in normal everyday motoring use. The MON number is measured under more severe conditions of higher engine speed and temperature, nearer to those you get on a highly-tuned competition engine being driven with a spot of verve. You usually find that, for the same fuel, the MON octane number is 10 to 11 points less than the RON number.

In the 1920s, the normal fuel at roadside pumps had a low RON rating of around 70, or even less. But the family car engines of the time had compression ratios around 5 or 6 to 1, and 2,500 rpm was considered quite a fast engine speed. As compression ratios, engine speeds and power outputs increased, detonation became rife and the petrol companies had to push up the octane rating of their fuels.

In the late 1920s it was discovered that adding lead compounds, usually tetraethyl lead and tetramethyl lead, to the fuel increased its resistance to detonation. These were the standard anti-knock additives used right up until a few years ago, and they are still used in today's leaded fuel, though the lead content has dropped from around 0.4 grams per litre to about 0.15 grams per litre and even lower because burnt lead compounds are nasty things to have around in the air we breathe.

Typically, today's leaded four-star fuel has an octane rating of 97 RON and 86 MON. Super-unleaded fuel is usually slightly higher, typically 98 RON and 87 MON. This is achieved partly during the refining process, but more by using additives known as oxygenates in place of lead compounds.

Yet another fuel factor affects the chance of detonation, the strength of the mixture. Broadly speaking, there is less chance of detonation with a rich mixture than with a weak one, partly because of the burning characteristics, but mainly because a weak mixture burns hotter. It's yet another factor to be taken into account by engine tuners.

The last point about fuels we want to touch on is clean burning. We hear a lot about detergent additives in fuel, but they're nothing new. Detergents have been added to fuel since the 1950s when their main purpose was to stop carburettors and fuel pumps gumming up. Better additives which helped to keep the inlet valve clean came in the 1970s, but even they aren't a patch on the new generation of detergent additives first introduced by Shell.

Their prime purpose is to keep the nozzles of fuel injectors clean, but they have the added bonus of going a long way to stop hard carbon forming in the combustion chamber and, as we saw earlier, lumps of hard, hot incandescent carbon can give us pre-ignition.

Back to practical tuning. Though most tuning is carried out on the inlet side of the engine, almost as important as getting more air and fuel mixture into the cylinders is getting the burnt gases out. Head porting can play an important part, but a good exhaust manifold and system is a huge help if you are going to reap the benefits of work done on the inlet side and the cylinder head.

The standard exhaust manifold on a non-GT Crossflow isn't the worst in the world, but it's a long way from being the best. You can do a lot to improve things by reducing the back pressure in the rest of the system which will give you more gain than just changing the manifold and leaving the rest of the system as standard. To get the most benefit you need to change both. There are two roads you can take on exhausts. You can buy an off-the-shelf system, either complete with manifold, or from the downpipe backwards, from someone like Piper, Janspeed or Peco, or you can go to a specialist builder and have a custom system made.

Because they have to be all things to all men, the proprietary manifolds have to be something of a compromise, but they are generally a good compromise. They aim at a wide range of tune from mild pep-up to fast road, and have to balance good slow running and a reasonably good torque low down the rev range against power higher up the rev range.

It says a lot for their careful design that they satisfy the vast majority of people who buy and fit them. If, however, you want to specify more closely where the extractor effect of the system is going to come in the rev range, you have to go to a specialist exhaust builder. Depending on the type of camshaft you have fitted, and the rest of your tuning modifications, he can tune the exhaust system to help give you almost anything from low-down torque of the wall-climbing variety to a very peaky power curve with all the power at the top end for racing. I'll deal more fully with manifolds and exhausts later on.

All this work at the top end of the engine presupposes that the bottom end is strong enough to take the increased power. Most engines, and Ford's Crossflow is no exception, have quite a lot in hand from a stress point of view, but it's little use spending a lot of money, time and effort getting more power from the intake, head, camshaft and exhaust if the bottom end of the engine has covered a hundred thousand miles and could really do with overhauling at standard power.

On the Crossflow, the camshaft lives downstairs in the block and the engine has to be almost stripped to change it, so there's no excuse for not checking the bottom end properly.

At the very least, you should fit new main and big-end shells, a higher pressure oil pump and a new timing chain. If the crank journals are worn oval, the shaft should be reground. Standard mains and big-end shells are quite adequate for mild tuning, but for more power you should fit a set of higher-duty lead-indium shells, often called 'competition' shells.

▼ Even if you aren't having the crankshaft reground, the least you should do when you uprate is to fit new lead-indium bearing shells, even on a low-mileage engine.

Balancing is very important. At the minimum, the crank and flywheel should be balanced. For higher power output, balancing should be extended to include the pistons and connecting rods. If you are going for top power you should consider things like duplex timing chains and roller valve rockers. Unlike the Pinto where, when you go for higher power, it's almost essential to have the flywheel double-dowelled to the end of the crankshaft, it isn't really necessary on the Crossflow short of full race tune. For circuit racing you should also look into having the sump baffled to combat centrifugal force flinging the oil away from the pump inlet.

All the points mentioned in this chapter will be dealt with in detail later in the book.

Though it has nothing to do with engine tuning, I would like to make a plea here about the rest of the car and safety. Before you start thinking about increasing the power of the engine, please make certain that the bodyshell of the car is absolutely structurally sound. Even mild corrosion, which may not be bad enough to fail an MoT test, can weaken structural parts enough to cause failure under the higher stresses of greater power.

For really high power you may need to change some of the spot welding to continuous seam welding, and possibly add strengthening plates to the body shell. Don't try to do this yourself unless you know exactly what you are doing. Inexpert or uninformed welding can raise stress points which weaken rather than strengthen the bodyshell. If you go racing or rallying you will have to meet a set of specifications about the strength of the bodyshell and internal roll cages.

The two other important areas are the suspension and the brakes. Many Fords fitted with Crossflow engines are older models, for example. The brakes and suspension were perfectly adequate for the cars and the road conditions at the time they were built. For higher power they need uprating, and there are numerous specialist firms who can advise you.

◀ *An uprated oil pump is a good investment on any tuned engine.*

▼ *Even if you are not uprating, and just rebuilding to standard, a new timing chain is always needed, and possibly new sprockets as well.*

◀ *With a performance camshaft you need a vernier chainwheel set, like this one from Kent Cams, to set the cam timing accurately.*

▶ Having the crankshaft and flywheel balanced will give your engine both a smoother tickover and a smoother delivery of power at high revs.

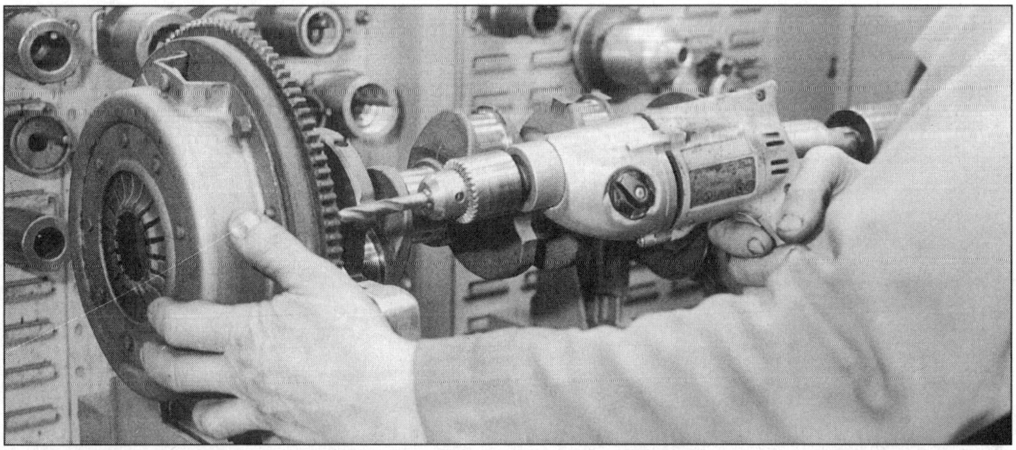

▶ If your distributor has seen many thousands of miles, either rebuild it or, as here, fit a new higher rated one from a tuning shop.

▶▶ A new set of top-quality plug leads is a good investment. The old ones fall off in efficiency so slowly that you may not realize how bad they are until you change.

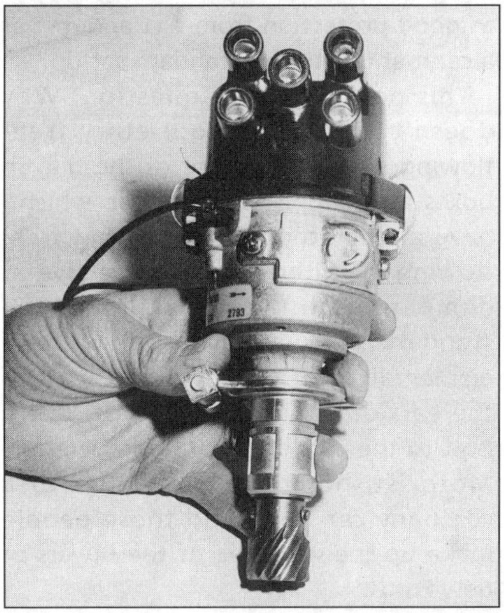

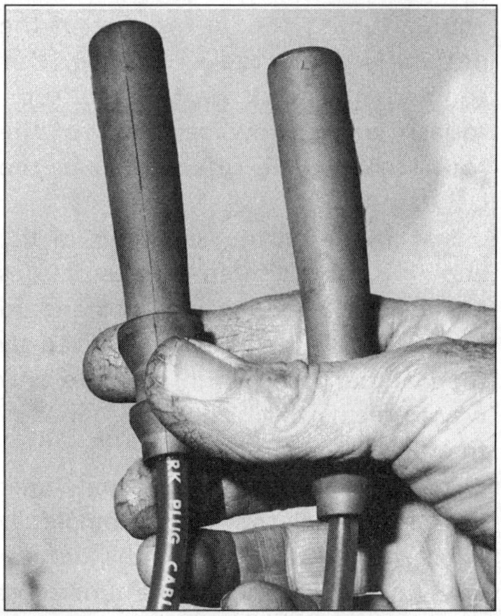

▶ For top power from a Crossflow you need to change to a pair of large side-draught carburettors, like these Weber DCOEs, on special manifolds.

Chapter 4

The intake side

Getting more air/fuel mixture into the cylinders is the principal aim of engine tuning and, in the case of the naturally aspirated Crossflow, this concerns the 'suck' part of the 'suck, squash, bang, blow' sequence of the four-stroke cycle referred to in the last chapter.

First in line on the 'suck' side of the engine is the air filter. For the reasons already explained, air flow in standard production filters is compromised by their all-purpose design, and the only way to rectify this is to fit a proprietary free-flow filter.

Free-flow is a little bit of a misnomer because the only way to get completely unrestricted flow is to leave the air filter out altogether. So, why don't tuners just throw away the air filters supplied by Ford and leave the top of the carburettor wide open?

Two reasons. The first is that no-one wants to suck dirt and grit into the engine along with the air, and the second is that you don't necessarily get more power with an open air intake.

Track and dynamometer tests by filter manufacturers and racing teams have shown that with a properly designed free-flowing air filter you don't get a loss of power. Indeed, in some cases the tests showed that you can get a slight increase in power, compared with an open intake, because of less turbulence around the mouth of the carburettor. Other tests have shown that a well-designed free-flowing air filter gives just as good protection from dirt and grit as a car manufacturer's standard set-up.

This poses another question. Why doesn't Ford fit a proprietary free-flowing air filter in place of the big tin boxes they supply. One answer, which I have already touched on, is that free-flowing air filters are less effective at dampening intake roar than the standard set-up of a tin box with an element inside it. This might be music to the ears of a tuning enthusiast, but it isn't to the ears of Mr and Mrs Average Motorist, nor to the manager of a company car fleet, and those people make up the vast bulk of the buyers of new Fords.

Another reason is that the paper element inside the standard air filter box is cheap. When it gets dirty you just throw it away and replace it with a new one. You could do the same with a free-flow filter element but it would prove expensive as you would have to change it every 12,000 miles or so. It's much cheaper to clean and re-oil it, which is something an enthusiast doesn't mind doing but which would, in the case of the average motorist, result in additional servicing costs or possibly become a job neglected.

There's yet another reason. As environmental laws have become tougher, engines have had to eat their

▶ *Free-flow filters come in shapes and sizes to fit just about every engine and installation.*

own fumes instead of pushing them out into the atmosphere. A convenient way to do this is to pipe the crankcase and rocker box breather pipes into the air filter box, often with an oil-trap filter. This is quite easy to do with a tin box filter casing, but not quite so easy with a free-flow pancake filter sitting on top of the carburettor.

There's also the question of icing up in winter. When the air rushing through the carburettor vaporizes the fuel there's a drop in temperature, and if the incoming air is anywhere near freezing the water vapour in it freezes out as ice crystals which block the carburettor jets.

Quite a few people have fitted free-flow air filters to find that they work fine in summer, but come the winter either the car won't start, or it starts and keeps petering out until the air under the bonnet gets warm. Even when under way in winter – cruising along a motorway, for instance – cold air flowing over the engine can suddenly cause things to freeze up. Although a short wait on the hard shoulder, while the heat from the exhaust manifold melts the ice in the carburettor, will solve the problem, it is not a desirable thing to happen.

There are several ways a car manufacturer gets round this. One is to make the air filter box, or its top, swivellable, so that in winter the intake pipe can be pushed round over the exhaust manifold to draw in warm air. Alternatively, the air filter box can have two intakes, one over the exhaust manifold and one well away from it, and a thermostatically controlled flap inside the box chooses one or the other according to the temperature.

A third method is to fit a coolant-heated manifold, which isn't in the best interests of power because for power you need oxygen and warm air is less dense, so it contains less oxygen per cubic centimetre than cold air. A slight loss in power is preferable to stopping, but a water-heated manifold has the disadvantage that it doesn't work till the coolant gets hot by which time most of the icing troubles are over. There is also the disadvantage that unless there is some way of shutting off the hot water to the manifold the slight loss of power continues in the summer when there is no need for extra heat.

It's true that a heated manifold sometimes helps fuel consumption and keeps the mixture more even by helping to prevent fuel droplets condensing out of the mixture on to the walls of the manifold, but it shouldn't be necessary with a well-designed and gas-flowed manifold. Much the same objections apply to hot-spot heating where the inlet manifold sits on top of the exhaust manifold, but this doesn't apply in the case of the Crossflow engine where the inlet and exhaust manifolds are on opposite sides.

If you're beginning to think that getting more air through the carburettor isn't quite so easy as might seem at first sight, you're dead right. Let's have a look at how we can overcome the snags and still get most of the advantages.

There are two basic types of free-flow air filter – cotton gauze and polyester foam. Well-known examples of gauze filters are K&N, Euro Web (which uses an element similar to K&N in a cheaper case) and JR filters which are imported from France. K&N filters are exclusively cotton gauze, but in filters from the other sources you will find either cotton or man-made fibre used. Among the polyester foam filters are Pipercross and Megaflow. All have to be sprayed with a special oil to get the best filtration.

Both types have their champions and both give you much more efficient air flow than the car manufacturer's standard air filter, so we'll look at some of the other claims.

Cotton gauze filters are robust and easy to clean and re-oil, but they cost more than foam filters. Provided you don't do anything silly, like cleaning them with harsh detergents or a wire brush, the makers give you what almost amounts to a lifetime's guarantee.

Foam filters can also be cleaned and re-oiled, though generally they don't last as long as cotton gauze because they are more delicate and tend to bruise. You can pay a lot of money for

◀ *A pair of Weber DCOE 40s with free-flow filters. Quite a tight fit in a Mk II Escort.*

▲ *K&N's 56-9070, a popular filter for fitting on a Weber DGAV or DGV.*

Because they are moulded and can be shaped to fit in awkward places, foam filters have the advantage in applications where room to put a decent-sized cotton gauze filter is restricted.

There is the question of fire. Makers of cotton gauze filters often stress that their filters won't catch fire – unless they're soaked in petrol – should your engine be unkind enough to spit flame back through the manifold. With an engine in good condition it shouldn't happen, but it might with a really wild-profile cam with lots of valve overlap if you try to start the engine with too much ignition advance and it kicks back. It could also happen if you get the ignition timing hopelessly out with even a mild cam.

▶ *Free-flow filters come with the element ready-oiled. It needs cleaning and re-oiling from time to time, but follow the maker's instructions closely.*

special foam filters, but for general applications they're less expensive than cotton gauze, and the need for occasional renewal can be offset against the cheaper initial cost.

Makers of foam filters claim that they get a better air flow, size-for-size, because they haven't got a top plate, and draw in air from the top as well as the sides. Gauze filter manufacturers counter this by saying that their specified filters for any engine are generously sized and never work anywhere near their full-flow capacity.

Polyester foam filters will burn and melt if you get a big flash of flame back through the manifold. In recent years the chances of this happening have been much reduced because foam filters are now made with several layers of foam – coarse outer layers which provide the main filtration and finer inner layers which are fire-retardant. They will still burn but they won't burst into flames anything like as easily as the early foam filters.

Makers of both types claim that their

particular product has better clog resistance; in other words, it doesn't need cleaning so often. I haven't carried out any 'A and B' type testing, and I don't know an independent body that has, so it's a question of listening to the salesmen and making up your own mind. One thing is clear, though, and that is that whichever type of free-flow air filter you choose you will get much more air through it than with Ford's standard air filter at any engine speed.

On, then, to the question of carburettor icing. This has always been something of a pain with free-flow filters. In winter, the air flowing into the carburettor can be near to freezing point yet still be loaded with water vapour, particularly in 'freezing fog' conditions. If the fuel which the air picks up from the carburettor jets is to vaporize, it needs heat – called the latent heat of vaporization. (You can feel this effect if you put some petrol on the back of your hand and hold it up in the wind. As the petrol vaporizes, your hand feels colder because the petrol has drawn the required heat from your hand.) In a very cold carburettor the only agency from which the fuel can get its heat is the air stream. This lowers the near-freezing temperature of the air to below freezing point, causing the water vapour to drop out as ice crystals.

I mentioned earlier that with the standard Ford tin-box air filter casing you can swing the inlet pipe over the exhaust manifold so that the incoming air is warmer. You can still do this, and get some of the benefits of a free-flow filter, if you fit a free-flowing gauze element into the standard Ford casing. Both K&N and JR make elements to fit almost every type of Ford filter casing. You don't quite get the best of both worlds, because a gauze element inside a tin box isn't as efficient as a bolt-on gauze or foam filter, but it will be much more efficient than the standard paper filter element.

◀ *Even if you don't change to a double-choke carburettor you can still benefit from a good filter. This one is made to fit the Weber ICH.*

▼ *It may not do anything for performance, but a filter top with the model name is an eye-catcher under the bonnet.*

Some people fit a bolt-on free-flow filter in the summer and change to a gauze element inside the standard casing for winter, and until a couple of years or so ago this was the only option. It still is if you choose a foam bolt-on filter, but with a K&N gauze filter there is a solution.

K&N has produced a metal funnel which clamps on the side of the filter between the top and bottom plates. Leading from this funnel is a flexible pipe which you can clamp over the exhaust manifold to suck in warm air, in much the same way as with the standard Ford filter. This wouldn't be a

lot of good if the rest of the element was open to suck in cold air, so K&N provide a couple of PVC strips with the kit. Depending on the height and diameter of the filter, you cut these strips to fit between the top and bottom plates, so you partially blank off the 'cold' side. When the kit first came out it didn't work all that well, but K&N say they've now modified it and it works fine.

You're bound to lose some of the potential air flow but, according to K&N, not as much as you might think. The company always sizes its filters on the generous side and has carried out

▲ *This is one way you can cater for the engine breather pipe that used to fit on the big tin-can filter.*

▶ *Another method of fitting the breather is to use an elbow, either in a grommet or held by a nut, in the base of the filter.*

▼*This is K&N's anti-icing kit. The trunking is led to a hot air point over the exhaust and the clear plastic strip is cut to blank off part of the element.*

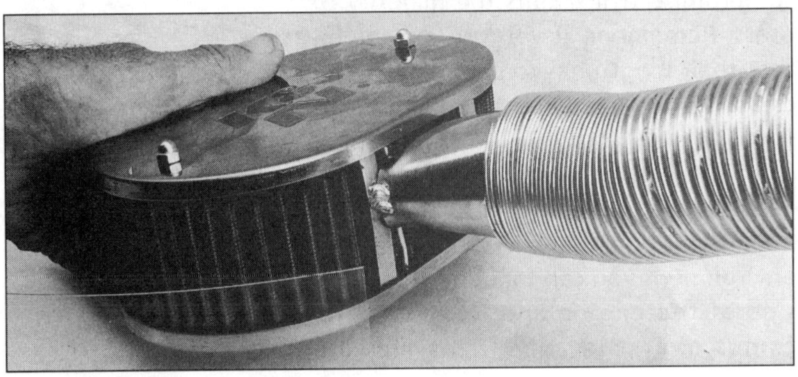

tests to show that, if you fit the kit properly, the drop in power on a normal fast-road tuned car is 'negligible'.

Both gauze and foam filters come in a huge variety of sizes. If you've decided to give your Crossflow only a very mild pep-up with a free-flow filter and a performance exhaust, there's no problem. You choose the standard filter from the maker's catalogue and bolt it on.

If, however, you've taken the tuning further with a gas-flowed head, larger valves and a performance camshaft, you have to make an estimate of what the bhp of your modified engine is going to be at the flywheel. Then you look in the maker's catalogue under 'power factor', or some similar heading, to give the minimum size filter which will suit that power. If in doubt, err on the generous side. Then you have to pick one that fits whatever carburettor you have chosen and which isn't too tall to fit under the bonnet. The catalogue will tell you what clearance you need over the carburettor for any given filter.

That clearance has to be increased if you want to fit a ram pipe or pipes (sometimes called air horns). Ram pipes are another way of increasing the air flow through the carburettor, but they don't work at the same efficiency over the whole engine speed range. When air is flowing through a pipe it doesn't flow in a constant even stream; it flows in pulses, with the frequency of the pulses depending on the speed of the air and the diameter of the pipe. The pressure and the flow at the outlet of the pipe rises and falls with each pulse.

Ram pipes are short lengths of pipe, sometimes with a venturi inside to speed the air flow, that fit on the intake of the carburettor. With a single-cylinder engine, the length and diameter of the pipe can be altered to tune the air pulses so that, at any given engine speed, a pulse arrives at the base of the carburettor at the same time that the inlet valve is open. You can't do this with a multi-cylinder engine and an inlet manifold, but the makers of ram pipes have carried out exhaustive combinations of calculations and testing to produce pipes which will give increased air flow into the cylinders over a fairly narrow band of engine rpm. Because ram pipes are used mainly on engines tuned for racing, most of them increase the air flow at the top end of the rpm range and do little or nothing at the lower and middle parts.

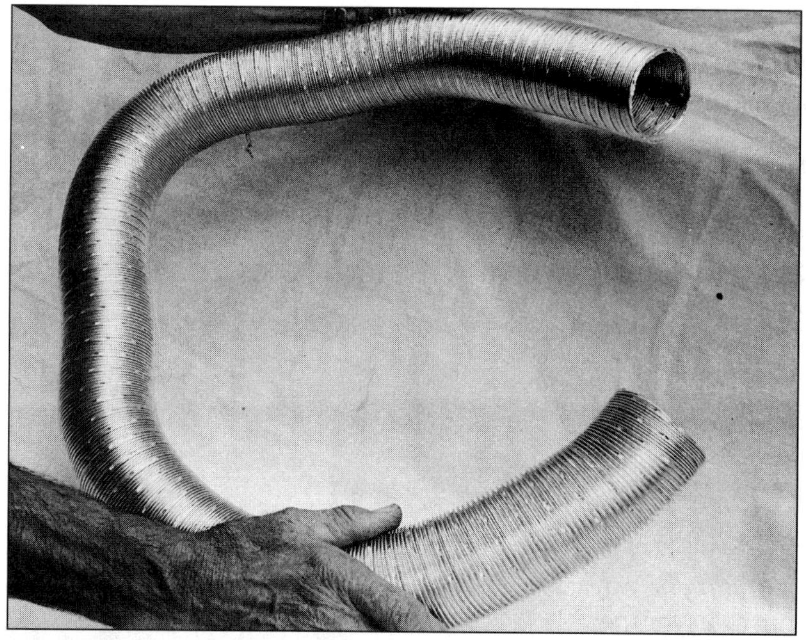

▲ *At the other end of the temperature scale, metal trunking like this can be used in summer to feed cold air to the filter.*

Some makers, however, produce a range of ram pipes which give the increase in the upper middle of the rpm range where it is of most benefit for a road-tuned engine. Ideally they would like to know what the flow characteristics and distribution along the manifold are, and the gas flow through the head at the rpm where you want the extra power. But having this tested and having ram pipes custom made can be a very expensive business. Like the makers of off-the-shelf performance exhausts, they settle on a compromise which suits the majority of users. Remember, if you use ram pipes, that their length has to be added to the clearance you need above the air filter.

You're still left with the rubber pipe from the engine breathing system that used to fit on the standard Ford air filter box. If you don't care a hang about the environment you can take this down the side of the engine and let it blow its fumes over the road. This may be

All that glisters is not gold. A cheap look-alike for a free-flow filter with a foam plastic element that clogged up quickly.

acceptable on a racing circuit, but for road use, as well as being anti-social, it may not be looked upon too kindly by the MoT tester. Alternatively, you could feed the pipe down into an oil trap so that air which goes out to the atmosphere is less oil laden, but it isn't the best solution and, in a few years time, may not be acceptable under EEC regulations even on older cars.

When you buy your free-flow filter, ask about a breather pipe adaptor plate which fits between the carburettor and the filter. To avoid sucking dirty engine oil into your nice new filter you will have to fit an oil separator valve-cum-filter in the breather pipe. The one which Ford uses on the carburetted XR3 Escort is quite an efficient one.

Alternatively, you can extend the breather pipe and run it in an upside-down U-bend alongside the filter so that most of the oil drops out of the fumes at the top of the bend and runs back to the rocker cover. It isn't as efficient as a proper oil trap and you will need to clean the filter element more often, but it's better than nothing. Once again, remember to add the thickness of any adaptor plate to the clearance you need over the top of the filter.

Carburettor

Now we come to changing the carburettor. As already mentioned, your Crossflow will originally have been fitted with either a Ford Motorcraft carburettor, a single-choke Weber, a double-choke Weber or Ford's VV (Variable Venturi). Both the Motorcraft and the single-choke Webers are good workmanlike carburettors but they don't lend themselves very easily to tuning, so you're better off changing them.

Ford's VV carburettor is another matter. It is difficult to beat for a combination of power and economy, and it will also adapt itself to quite a wide variation in air flow. Unfortunately, when it gets a little old, it has a habit of going off song, and when it does it can be a real pig. The only thing to do is strip and completely rebuild it, and it isn't the easiest carburettor to strip, rebuild and set-up by quite a wide margin. If you can find someone who really knows their stuff about tuning a VV then by all means stay with it. If not, I would advise changing it.

On tuned Crossflows most people go for the Weber twin-choke. They are popular because they are efficient, they hold their tune and, despite being somewhat complicated to look at, they are easy to overhaul. There are agents all over the country who either have overhaul kits in stock or can order them for you quickly, even for older models.

It's worthwhile here saying a few words about Weber's coding system.

Take some popular examples – a 32 DGV, a 32/36 DGAV and a 40 DCOE. So far as the letters go, we need concern ourselves only with the D, the A (or lack of an A) and the O. The D stands for double choke – choke in the sense of barrel – and A stands for automatic choke – choke in the sense of strangler. These are both downdraught carburettors with the inlets at the top. On the DCOE, the O stands for Orizontali, or horizontal inlet – what many people refer to as sidedraught. The other letters mean something only to the Weber design office.

Some people get misled by the word choke. The popular use of the word means either a strangler butterfly or an enrichment device to help cold starting, and carburettor manufacturers also use it for this. However, they also use it for the venturi-shaped inlet to the carburettor. So, when we talk about a double-choke carburettor we mean one with two inlet barrels. It will also have twin strangler butterflies, but we refer to these as chokes only when we are talking about manual or automatic choke, in other words, manual or automatic operation of the cold-start strangler butterflies.

The other letters refer sometimes to a family of carburettor designs and sometimes to the make of car for which the carburettor was originally designed, even though it has since been modified to suit other makes. For example, an L in the code of a DMTL is popularly supposed to indicate that the original production was designed to fit Leyland Group cars. Similarly, the R in a DGR is popularly taken to mean that the original was designed for Ladas – R standing for Russian.

The figures in the code stand for the choke diameters in millimetres. Here, we're talking about the inlet barrel, not the strangler butterfly. Where you have one pair of figures, as with a 32 DGV, the D means double choke and the 32 means that each choke is 32mm diameter. On something like the 32/36 DGV you have two chokes, one 32mm diameter and one 36mm diameter.

The two most popular models for road tuning a Crossflow are the DGV and the DGAV. The only difference between the two is the manual or automatic choke, and this is entirely a matter of preference. Some people are prejudiced against automatic chokes and won't have them at any price. Others, possibly because they know how to set them, love them because if, for example, they are tuning a Mk II Escort originally fitted with a 32/36 DGAV they can overhaul the carburettor themselves and not have to bother with fitting a linkage for a manual choke.

It's worth mentioning that you can get a kit from Weber to convert from automatic choke to manual choke, and this is very helpful if you like a manual choke and want to save yourself quite a lot of money by not having to buy a new carburettor.

Let's suppose, for example, that you have an 1100 or 1300 Crossflow with a Ford Motorcraft carburettor and you want to change to a double-choke Weber. There are literally hundreds of 32/36 DGAV carburettors sitting on pre-Mk III 1600 Cortinas with Crossflow engines in breakers' yards. The 32 DGV was fitted to the Escort Mk II 1600 GT and Ghia and to the Escort Mk II 1100 GT and 1300 GT. The equivalent Mk I Escorts had either a DFE or DFM carburettor. Overhaul kits are available from Weber for all these and all can be rejetted and tuned to suit Crossflow engines modified to fast road use. I deal with overhauling a 32/36 DGAV in a later chapter.

Suitably jetted, these carburettors are quite big enough to cope with the mixture demands of any Crossflow tuned for normal fairly fast road work. You will need to go to a 40 DCOE, 45 DCOE, or a pair of DCOEs, only if you are going in for higher power. In that case you won't be modifying your engine at home, you'll be using a specialist tuner who will have prepared the engine to use carburettors with that rate of mixture flow.

The Manifold

Next, on the inlet side, is the manifold. From a bolting-on point of view, any inlet manifold from a Ford Crossflow will fit on to any other Ford Crossflow. Where they differ is in the internal design, the size of the bores and outlet ports and the mounting flange for the carburettor.

This opens up a big field for the home tuner of a Crossflow. Some of the inlet manifolds on standard 'cooking' Crossflows, particularly the ones with a single-choke carburettor, were very pedestrian affairs, suited to the limited power output of the engine in standard tune but of little use when you're looking for uprating the power, because they are far too restrictive.

If you want to keep costs down and stay with standard Ford manifolds that you can buy off the shelf or from a breakers' yard, the ones to go for are the ones from the 1300 GT Escort Mk I, the 1100 or 1300 Escort GT Mk II, the 1600 Escort GT, Sport or Ghia or the pre-Mk III, pre-overhead cam, Cortina 1600 GT with a Crossflow engine. All these were fitted as standard with a double-choke Weber and all have quite reasonable flow characteristics, even though not as good as a proprietary performance manifold.

I said that one area where the manifolds differed was in the size of the bore and outlet ports. Too few home-tuners understand the importance of matching the outlet ports on the inlet manifold to the inlet ports on the cylinder head. If you bolt a manifold from a 1600 onto the head of a 1300, the incoming gas will meet a ledge where the port in the head is smaller than the port in the manifold. This will set up so much turbulence that the flow will be considerably worse than with the original manifold. You have to modify either the head, the manifold or both to match each other.

This brings us to the already referred to golden rule of intake-side tuning – matching the ports. Matching the ports doesn't mean making them exactly the same size. You get the best results if the port in the manifold is slightly smaller than the port in the head. There are two reasons for this. The first is that it avoids any possibility of the incoming gas meeting a ledge that will cause turbulence. The second is that, at high revs and certain valve overlap conditions there can be short periods of slight back-flow of gas back into the inlet manifold.

The back-flow tends to travel along the outside of the head ports. If it comes to a slight ledge when it reaches the inlet manifold it will be deflected back into the main stream of incoming gas. There will be a small amount of undesirable turbulence in the cylinder head port, depending on the amount of back-flow, but its effect on the power and the smooth acceleration of the engine will be far less than if the back-flow worked its way up the manifold to the carburettor. The difference in size doesn't need to be much, about 1.5mm all round is ample.

Together with matching the head and manifold goes matching the manifold gasket. Always match the holes in the gasket to the larger ports – the ones in the cylinder head. Now a word or two about how to match them on the home bench. First, matching the manifold.

You may have had the head from your engine gas-flowed by a tuning firm; you may have bought one of their heads already gas-flowed; or you may have gas-flowed it yourself. I deal with gas-flowing a head in a later chapter, but for the purposes of matching the manifold the port size and shape in the head will already have been decided. They will be slightly larger than standard so you have to enlarge the ports in the manifold to match.

When you offer the manifold up to the ports on the cylinder head you may find that the holes where the fixing bolts go are large enough to let the manifold move around slightly. If you match the ports in the manifold accurately to the ports in the head in one position the chances are that you will overlap one or more of the head ports when it moves, and create an undesirable ledge.

The way to overcome this is to shape the ports in the manifold so that however far it moves on the fixing bolts, it does not allow any port on the head to be uncovered. Alternatively, you could enlarge the ports on the head a little. Neither is the ideal situation from a gas-flow point of view because the port in the manifold will be slightly smaller than it need be, but the effect on power will be very small, and certainly not noticeable in practice on a road car.

You can't, of course, see the ports when the manifold is bolted on to the head, so you have to have a template from which to work on the manifold. The easiest way to make a template is to get a spare manifold gasket – a used one will do provided it's clean and undamaged. Attach it to the manifold with a non-setting jointing compound and then paint the port face of the cylinder head with engineer's blue.

With the demise in the average garage of the fitter, a man who shaped things to fit as opposed to a mechanic who bolts on replacement parts, engineer's blue is not so easy to get as it used to be. You may have to go to a firm that supplies machine shops. It looks like a thin blue paint and it stays wet long enough to leave a blue mark on anything that's pressed against it.

When you bolt the manifold and gasket to the blued face of the head you will get the outline of the head ports in blue on the gasket. Go round these outlines with a scriber before you smudge the outlines. Now you can cut the port holes in the gasket to the exact size of the head ports and use it as a template when you reshape the manifold ports.

You can shape the ports in an aluminium manifold with power-driven riffler files fitted in your electric drill. You can buy these in engineering tool shops, sometimes in boxed sets and sometimes singly. For matching a manifold you want parallel ones with a working length about two inches. Choose the diameter to fit comfortably in smallest radius of the ports in the cylinder head. If there is a choice, choose ones intended for working on aluminium, and choose a medium to fine cut. For working on cast iron you need riffler stones and, ideally, a very high speed hand grinder in which to fit them. You can do it using the fastest speed on your electric drill, but it takes longer.

Scribe the face of the manifold from the gasket template and enlarge the ports with a gentle taper going back into the manifold about three quarters of an inch rather than an abrupt chamfer. Test with the template frequently. It's much easier to take more metal off than it is to put metal back. You can also use your template for cutting the new manifold gasket to fit.

Chapter 5

The exhaust side

In Chapter 3 I said that almost as important as getting more mixture through the intake side of the engine is getting the burnt gases out of the other side. I used the word almost, because many tuners hold the view that if, on the exhaust side, you can achieve 80 per cent of the gas flow on the inlet side you are doing well. If your budget is tight they will advise you to spend your money on the inlet side of the engine and leave the exhaust system until later. Part of the reasoning is that it's easier to change the exhaust at a later date than make alterations to the inlet side.

Naturally, makers of proprietary performance exhausts won't necessarily agree with them and, once you have chosen your camshaft to give you the emphasis on power increase over the rev range where you want it, an exhaust tuned to match will give you the greatest benefit from the rest of the money you have spent.

Tuning an exhaust starts with the manifold and, with the exception of its GT engines, the standard Ford exhaust manifolds on the Crossflows are poor. The burnt gases come out of the cylinders because the piston is pushing them and the exhaust system is open at the back end, but the design does little or nothing to help extract the gases. Indeed, in many cases it positively hinders them.

Standard systems can give back-pressure which results in backflow so that, over some rev bands, the charge in the cylinder consists of 85 per cent new mixture and 15 per cent burnt gases. Quite a high proportion of these burnt gases have been out of the cylinder and have come back in again.

It's easy to pour scorn on Ford but, as with the engine designer, the person who designed the exhaust manifold and system had to produce something which was reasonably efficient, cheap to make and quiet. It's a sad but inescapable fact that very quiet exhausts don't equate with very efficient ones.

When you're looking at brochures to choose a better exhaust system, claims of so many per cent increase in power don't mean an awful lot by themselves. You want to know whereabouts in the rev range that increase in power comes and what effect the new system is going to have on the general shape of the engine's power and torque curves.

The designer of a performance exhaust system is faced with a whole host of conflicting requirements. On a fast road car the driver wants to feel the extra performance in mixed traffic driving, with a brisk turn of acceleration from 30 mph without having to row the car along on the gear lever. He wants good flexibility, which means he wants good torque low down. He also wants extra power at high revs for safe overtaking when he does drop down a gear. But he still wants the car to be smooth and quiet at 70 mph motorway cruising, and he expects it to tick over at about 800 to 900 rpm without lumping or shaking.

The exhaust designer has to take into account the characteristics of the tuned engine. No exhaust system made will give you more power than the engine is capable of producing. Nor will it give you low-speed torque if the rest of the tuning on the engine has been designed to put all the torque and power at the top end for racing.

It's a tribute to the mass-producers of performance exhaust systems such as Janspeed, Peco and Jetex that they produce systems which help the engine, or at least don't hinder it, to meet most of the requirements of the average home tuner, and at a very reasonable price. To do this they have to aim their systems to give maximum help where most people want it. With an engine tuned for fast road work this in the middle and lower-upper rpm range, if we can use such a phrase. In figures most aim, with an off-the-shelf manifold, to give the maximum help from about 3,500 rpm upwards. The mass-producers also have to concentrate on the popular models where they expect to sell reasonably large numbers. Fortunately, these include almost all the Fords fitted with a Crossflow, so you should be able to obtain an efficient performance system off the shelf.

If you are fitting your tuned engine into a kit car, the mass-producers will try to help you by adapting their standard systems but, if they just won't fit, you will have to go to one of the custom exhaust builders who will design and build a system for you. If you fancy your prowess with a welding torch or MIG welder, you could have a shot at designing and making your own – or at least your own system from the end of the primary pipes backwards. This is not quite such a daunting prospect as it might seem, as some custom makers, like Jeff Howe Exhausts at West Kingsdown in Kent, can offer you piping, bends, collectors and absorption boxes in a variety of sizes, and they will give you advice on the sort of pipe diameters and lengths you should aim at.

Before you can think about making your own exhaust system or, indeed, before you go to talk to a custom exhaust builder, it's as well to have some idea of the theory behind performance exhaust design. When I say theory, I don't mean that you have to study the theory of gas flow through pipes, and pulse frequencies. The mathematics of this can be daunting in the extreme. Most custom exhaust builders designed their original systems for a particular model using the standard formulas, but when they have to adapt they usually work on a combination of well-tried practice and experience with a minimum of mathematics.

Before you start designing a system or before you go to talk to a custom builder you have to set out a number of requirements, such as how much power you are aiming for, what sort of gas flow though the engine you expect, what the safe maximum revs of your tuned engine will be, how many gears you've got and their overall ratios, the tyre size you're using, and whereabouts in the rev range you want the maximum torque and maximum brake horsepower.

For something like circuit racing it's usual to have the peak power right up at the top end, with the maximum torque not far below – even if this means that the power curve is very steep and peaked and that the engine has little flexibility below about 3,000 or even 4,000 rpm. The exhaust system would match up with quite a wild camshaft and close ratio gears, and the driver would accept that he has to change gear frequently to keep the power turned on.

For fast road work, with long-distance cruising, you would want the power curve to be less peaked and less steep lower down the rev scale, equating with a rev band corresponding to about 60 to 80 mph in top gear without falling off too sharply, so that it's still good at higher revs in the intermediate gears.

If, on the other hand, your fast road car has to double as a family saloon, possibly even towing a caravan on holidays, you don't want to put a racehorse between shafts. You have to aim at a compromise with a still broader top to the power curve and plenty of torque in the medium rev range.

Except for the out-and-out racing engine, you have to accept some degree of compromise. As a general rule, the more maximum power you extract from an engine, the more peaky your power curve is going to be. For a broader power spread you have to accept a lower maximum, which is why choosing the exhaust system has to be part of the whole tuning plan, tied in with the carburettor, camshaft, cylinder head gas-flowing and the rest of the modifications.

Whichever type of power spread you set out to achieve, the maximum power output should always be set below the maximum safe revs of the engine. On a highly tuned engine it may be only 500 rpm below, but for a fast road car 800 to 1000 rpm is a safer margin. The exhaust system should be matched to give the maximum help to the outgoing gases at the same rpm.

Exhaust system design starts at the manifold where separate primary pipes – the colloquial bunch of bananas – will always be the most efficient when you are looking for top power. The primary pipes flow into a junction box, usually called a collector, shaped rather like a funnel with either four or two inlets and one outlet. The taper angle of the collector can also have an effect on gas flow, but most specialist exhaust builders have dies for stamping them out and may have to settle on an angle which is a compromise between the best gas flow and physically getting the system in the engine bay.

Once the specialist exhaust designer knows the capacity of the engine and its gas flow rate, and the revs at which you want the maximum extraction effect, there are several formulas he can use to determine the length of the primary pipes.

One quite widely used formula is:

$$L = \frac{5100 \times ED}{RPM \times 6}$$

Where: L = Length of primary pipe, in inches, measured from the exhaust valve head

ED = Exhaust valve duration in degrees from the point of opening to TDC

RPM = Estimated revs at which maximum power will be achieved, minus 500

Because L is taken from the exhaust valve head, the length of the exhaust port has to be deducted.

After arriving at a primary pipe length, the exhaust designer then wants to know what diameter to make it. Another popular formula is:

1. Take the capacity of one cylinder in cubic centimetres and double it.
2. Divide this by 16.4 to bring it to cubic inches.
3. Divide by the calculated primary pipe length.
4. Divide by π (or 3.1416).
5. Find the square root.
6. Multiply by 2 and add 0.120.

This gives a theoretical diameter for the pipe in inches, but it will be too small as the formula assumes a perfectly straight smooth pipe. To allow for air drag in the bends, the designer will increase the cross-section area of the pipe by 10 to 15 per cent. This will probably give him a non-standard pipe size, so he takes the next larger standard size.

With a four-into-two-into-one system, the length of the second pair of pipes in the manifold can be the same as the primaries, or a multiple of the length, with the diameter calculated the same way but, as each pipe is dealing with two cylinders, the capacity of two cylinders is taken.

Ideally, the length of the rest of the system should be a multiple of the length of the primaries and, as it is dealing with four cylinders, the diameter is calculated using the capacity of all four cylinders.

It all sounds very alchemical, rather like one of those quiz questions that end with 'take away the number you first thought of' but, when you're designing a system with bends and absorption boxes in it, and theory deals with straight pipes – let alone trying to make the system efficient over a wide rev range – science tends to take second place to try-it-and-see experience.

The ideal is to have all the primary pipes the same length, and although, because of restricting room in the engine bay, this is not always possible, they should be as near the same length as can be achieved. The fewer bends there are in the primary pipes the better and even if, again because of space restriction, some primaries have to have more bends than others, the bends should be the largest radius that can be accommodated.

The bore of the primary pipes is the important factor in determining whereabouts in the rev range you are going to get the maximum torque. The maker will start with a rough design of a primary bunch that he can physically get into the engine bay, and further calculations might show that the maximum torque comes at, say, 6,000 rpm.

Equating this with road speed in the various gears, he might decide that this is a little on the high side and want it pulled down to, say, 5,000 rpm. The peak torque can be moved down, albeit with a slight loss of maximum brake horsepower, by reducing the primary pipe diameter. As a rough rule of thumb for an engine the size of the Crossflow, reducing the diameter of the primaries by an eighth of an inch will move the peak torque down by around 650 to 800 rpm. Conversely, increasing the diameter of the primaries will raise the engine speed at which maximum torque occurs by approximately the same amount for each eighth of an inch. Once he knows he stands a good chance of physically getting the bunch into the engine bay, the designer can calculate it more accurately.

Changing the length of the primaries, within reason, has less effect on the position of the maximum torque. It's greatest effect is to alter the shape of the engine's power curve in relation to the rpm for maximum torque – rocking the power curve about the torque peak as one designer puts it.

Adding to the length of the primary pipes will lift the mid to lower rpm power with a reduction of power at maximum rpm. Shorter primaries will give a boost to the power at the top of the rpm range at the expense of power at the mid and lower rpm. Even then, the designer might have to accept a compromise because of physical limitations inside the engine bay.

The system most frequently used with racing engines is a four-into-one-system,

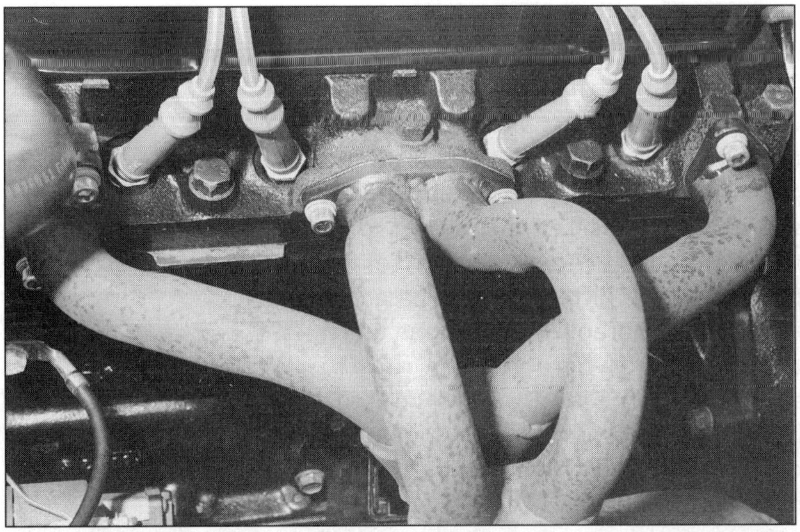

▲ *A typical 'bunch of bananas' performance exhaust manifold.*

where the four primaries come down to a single collector. But, for fast road work it is generally better to use a four-into-two-into-one system. In this system the primaries are in pairs with a collector for each pair. The outlets from these go to another collector which brings them to a single secondary pipe.

Four-into-two-into-one systems sacrifice power at the top end of the rev range to give better driveability lower down. A four-into-one with short primaries on an engine with a wild camshaft can be tiring to drive in traffic because the engine will feel lumpy at lower revs. The tickover will be something like 1,500 or even 2,000 rpm, and you might have to use the clutch most of the time to take it slowly through traffic.

With a four-into-two-into-one system, the diameter of the primaries has more effect on shape of the power curve than with a four-into-one. Increasing the diameter of the pipes will boost the torque and power at the top end of the curve without destroying the driveability too much. Conversely, reducing the diameter of the primaries will raise the shape of the torque and power curves lower down. The engine will tick over more smoothly and the low speed acceleration will be better. But, when the revs increase to near maximum the gases just won't get out of the combustion chambers fast enough, so the acceleration drops off earlier. Getting the last few mph maximum speed out of the car could take ages. Primary pipe design really is a question of swings and roundabouts.

The importance of thinking about tuning the engine and exhaust as a whole rather than two separate units is emphasized when you come to match the exhaust manifold up to the ports on the cylinder head. If you are buying a proprietary performance exhaust system you must know the diameter of the primary pipes at the top flange before you ask anyone to gas-flow and port the head. Alternatively, if you are going for larger diameter valves and larger diameter exhaust ports, talk with the head-tuner about how big the ports are going to be and, if you are using a proprietary manifold, get in touch with the maker to see if it can be ordered with larger bore primaries. Some cater for this, others don't – it depends how popular the alteration is on a particular engine.

Having the exhaust ports in the head larger than the diameter of the exhaust primaries would mean that the gases hit a circular wall as they try to flow out of the head port. This would set up turbulence which might well destroy all the benefits of increased gas flow through the head. Gas-flow figures for a head determined on a flow test bench are often measured with an open port, not with an exhaust manifold hanging on it.

Some exhaust makers match the internal diameter of the primaries exactly, or as near as they can get it, to the diameter of the exhaust ports on the head, but I prefer the internal diameter of the primaries to be about an eighth of an inch larger. Theory says that gases slow down when they go from a smaller to a larger diameter pipe, but any slight slowing down is much to be preferred to turbulence. The other reason is that head porting is done by hand, and the chances of the person grinding the port getting it truly circular are small. It's asking too much. A slightly larger diameter primary can cope with a small amount of ovality.

In our view, having the diameter of the primaries slightly larger than the ports also has a beneficial effect on any backflow from back pressure in the rest of the system. Gases which are back-flowing flow much more slowly than the main stream of gas on its way out, so they creep up the wall of the pipe. When they come to the sixteenth of an inch wall where the port in the head is slightly smaller than the diameter of the pipe they are deflected back into the main stream of gas and flow down again instead of finding their way into the combustion chamber.

It is important to remember that the standard exhaust manifold gasket as supplied by Ford will have to be modified if the head ports have been enlarged, or the gasket will provide a serious restriction to the flow. Not only will you lose power, there is a distinct probability that you will cause the engine to overheat. You also have to be prepared for the holes in the gasket not to match up exactly to the holes in the manifold as well as to the bolt holes. Bolt the gasket to the manifold with a couple of large washers to check the fit.

The usual practice is to make the holes in the gasket match the holes in the primary pipes rather than the ports in the head. That way, if the primaries are slightly larger, the gasket is trapped between two tightly bolted faces with only the edge exposed, instead of having a sixteenth of an inch round each port exposed to the hot gases where the chance of burning away is greater.

The length of the pipe from the last collector to the first silencer box – or absorption box, as exhaust designers prefer to call it – should in theory be the same length as the primaries or a multiple of them, but different designers have different ideas based on experience. Moreover, limitations under the car about where you can run the secondary pipe and where you can position the absorption box sometimes cause calculated theory to be thrown out of the window.

One well-known and successful exhaust designer, Mark Gosling who runs TTCM in Macclesfield, reckons that many secondary pipes are not long enough to get the best performance out of the primary system. As a rule of thumb, based on years of trial and testing, he has a method by which he adds three inches to the length of the primary pipes and then multiplies by two and a half. He agrees that it doesn't sound very scientific, but it gives him good results when tested on rolling roads.

He says that you can shorten the length of the secondary pipe within reason without upsetting things, but he has found that if you do it is necessary to increase its diameter. Once you get past the first absorption box the diameter of pipes in the system becomes less critical provided they are a large enough diameter to avoid back pressure. Quite a few people go for 'big-bore' rear pipes to avoid any chance of back pressure and, let's face it, because they look good and have a nice deep powerful sound.

There are on the market several 'extractor' additions which can be fitted to a tail pipe, and new ones appear from time to time. Some of them work on the principle of a collector funnel round the outside of the tailpipe which, because of its decreasing diameter, increases the speed of the air passing through the device from the normal passage of the car through the air. The claim is that this faster rush of air round the outside of the tail pipe has an extractor effect on the exhaust gases and thus increases the power of the engine.

I haven't carried out any comparison tests with and without one of these extractors, but the opinion of specialist exhaust builders is non-committal. In theory they help the exhaust gases out, thereby avoiding back pressure, and the theory holds water, but most specialists seem to be of the opinion that, though they might possibly have a beneficial effect with a car maker's standard exhaust, they are not needed on a properly designed performance system unless the system has been designed for them, in which case they will be built in.

However, Piper, Janspeed and others have been doing a lot of experimenting and testing with tail units consisting of a specially designed absorption box and a large bore outlet. I'm not familiar with the theory, and they're fairly close about it, but when tested on a rolling road they seem to give more power over quite a wide rev range.

▼ *Bending exhaust pipes at Piper RS.*

What is worth considering more deeply than many people do, is the design and material of the absorption boxes. All absorption boxes work on the same principle and consist of a central through tube which is perforated. Outside this is a packing of sound-absorbing material and, finally, the outer casing.

The better, and more expensive, designs have a layer of stainless steel wool next to the perforated tube and then a layer of basalt wool. In the cheaper ones, which look much the same, the only packing inside is glass fibre wool. When they are new there may be little to choose between the two, but when they get older the glass fibre wool in the cheaper ones gradually burns down to small globules which fall through the perforated tube and are blown out of the back.

I watched one of these cheap absorption boxes, about three years old, being sawn in half. From the outside it still looked fine, but the packing inside had deteriorated down to a few shreds of glass fibre wool and half a handful of melted globules, some stuck together. Far from being an absorption noise reducing box it was a resonating box which actually increased the noise level and was causing a resonant drumming inside the car.

Should you go for a system in mild steel, or is it worthwhile paying quite a lot extra for stainless steel? It depends on how long you want the system to last bearing in mind its initial cost. So far as primary pipes are concerned, stainless steel has very little to offer over a good mild steel. Though it is the hottest part of the system – indeed because it is the hottest – it corrodes far less than the later and cooler parts where the acid fumes and water vapour in the exhaust gas cool down and drop out as dilute liquid acid. Stainless steel will resist this corrosion from inside, as well as corrosion from outside, far longer than mild steel, no matter how many coatings it has on it.

Remember, though, that prices for stainless steel systems vary quite considerably. They may be made of the same quality stainless steel, but the cheaper ones are usually of a thinner gauge material which is less costly in the first place and easier to bend and shape. The thinner material also resonates quite badly in some cases and, if the natural resonating frequency of the car body shell coincides with that of the exhaust system you can get a most uncomfortable drumming inside the car at some engine rpm.

All performance exhaust systems will be noisier than the car maker's standard system, where quietness is a major factor in the design – more important in many cases than efficiency. To be reasonably quiet a good performance system, mild steel or stainless steel, should be made from at least 16 swg thickness material, and the casings of the absorption boxes should be seam welded, not folded and crimped. In stainless steel, such a system should last the life of the car.

When catalytic converters were first introduced and became mandatory on new cars, the idea of having to fit one in a high performance system filled many specialist builders with foreboding. However, since then a lot of work has been going on in the design of more efficient catalytic converters – efficiency from a performance point of view rather than from a converting point of view. At the time of writing there isn't, as far as I know, a catalytic converter that is as efficient from a gas flow point of view as a well-designed absorption box, though some of the better ones are starting to run it close. Before long, no doubt, there will be performance catalytic converters which detract little, if anything, from the efficiency of the system.

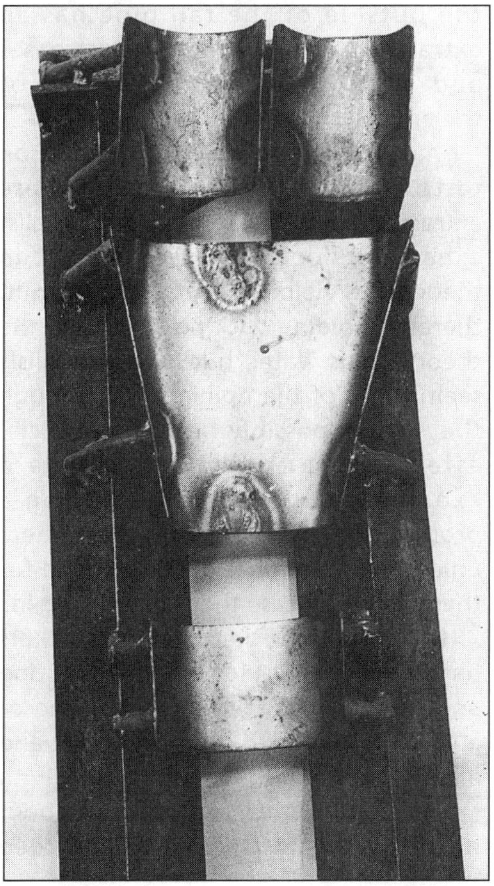

◀ *At Piper, exhaust manifolds are assembled on jigs like this.*

▼ *A complete performance system from downpipe to tailpipe.*

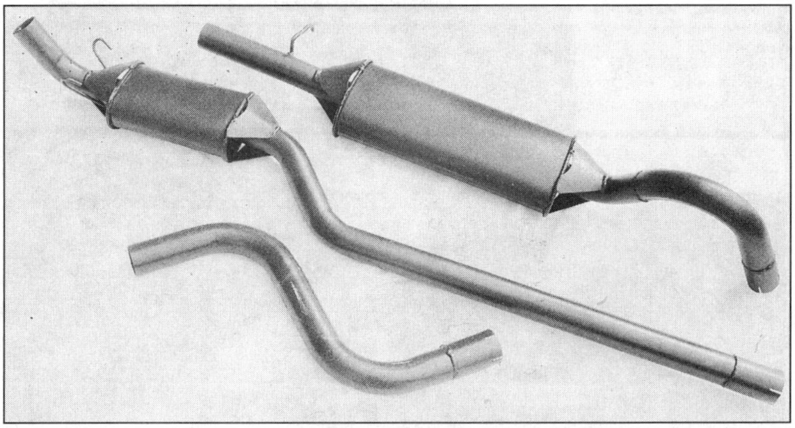

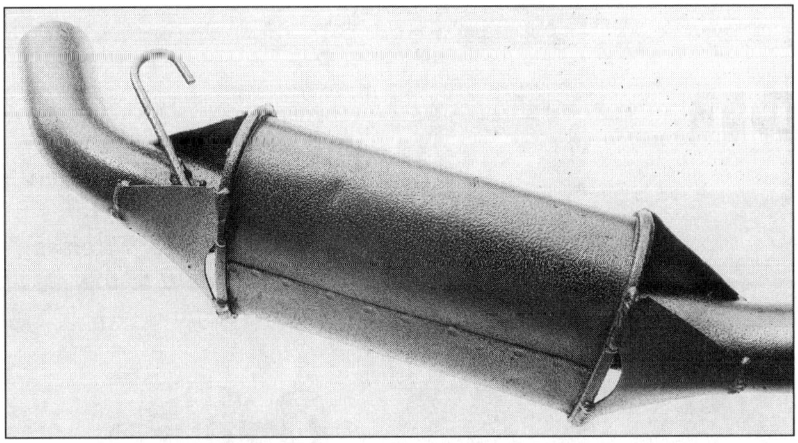

▲ *Bracing pieces to prevent cracking at the ends of expansion boxes are one sign of a good quality exhaust.*

◀ *A 'straight-through' expansion box has a tube of perforated metal surrounded by a silencing chamber packed with sound-deadening material.*

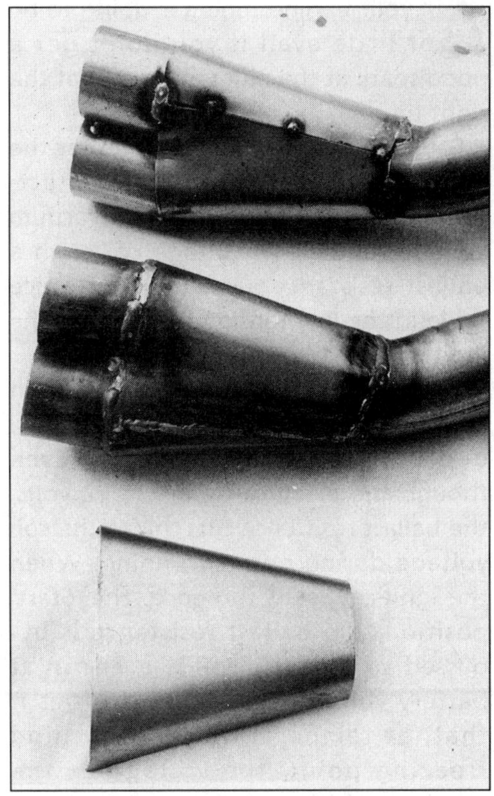

▶ *Stages in making a collector. It is tack-welded in the jig before being seam-welded. On the right is a blank half of a collector plate before shaping.*

▼ *If you don't have a pipe bender you can make up bends from small sections tack-welded for fitting, then seam-weld when the fit is correct.*

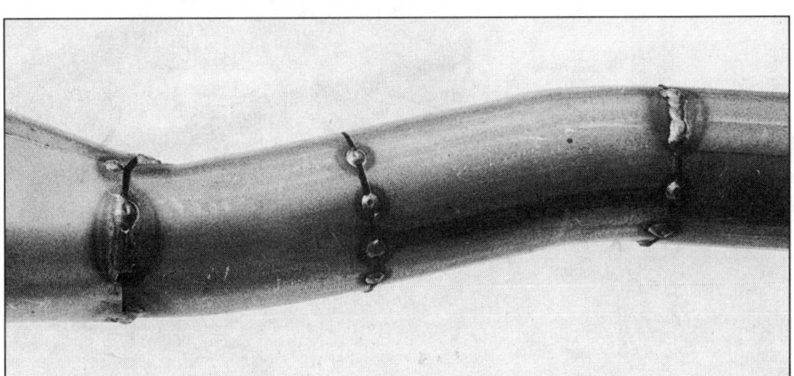

Chapter 6

The ignition system

▼ *An exploded view of the Motorcraft distributor fitted as standard to the Crossflow.*

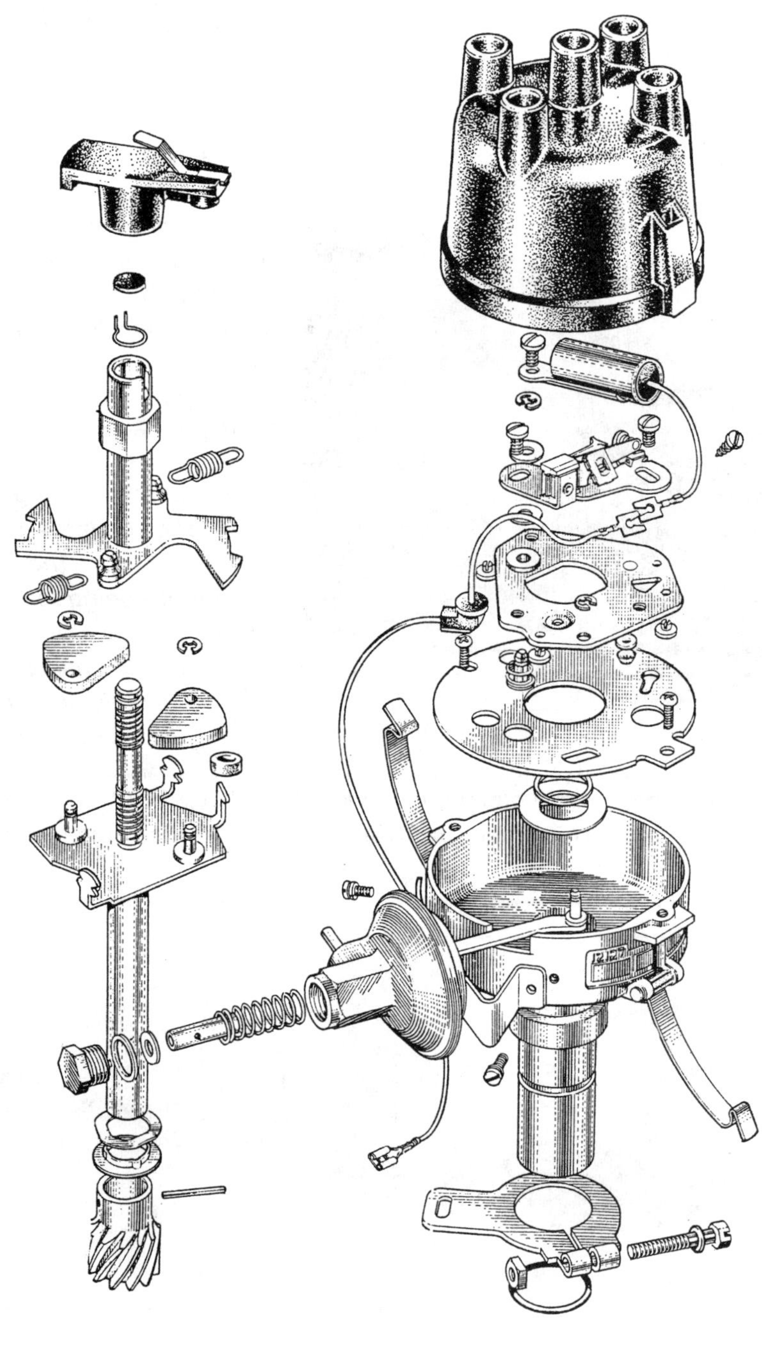

All your engine tuning is going to be of little avail if you don't get a good spark at the plugs and get it at the right time.

Ford's standard ignition system for the Crossflow uses conventional contact-breaker points, mechanical and vacuum advance and a low-voltage coil with a ballast resistance, usually a resistance cable in the low-tension circuit from the ignition switch to the coil.

The coil is designed to run continuously at either 6 or 7.2 volts, depending on the type of coil. Even though the electrical system is 12 volts, the ballast resistance cuts this to the coil voltage during normal running. When the ignition key is turned to the 'Start' position, the ballast resistance is by-passed so that the coil is over-run at battery voltage. The reason for this is that, at temperatures approaching freezing point, the voltage at the terminals of even a fully-charged 12-volt battery may well be down to 10 or 11 volts. The starter motor draws a very heavy current to crank over the cold engine and this causes a further voltage drop so the by-passed voltage at the low tension terminals of the coil may be down to 8 or 9 volts.

Using a lower voltage volt coil means that it will give a good spark under these conditions. On a hot day, a fully-charged battery may deliver 10 to 11 volts when cranking a warm engine, but

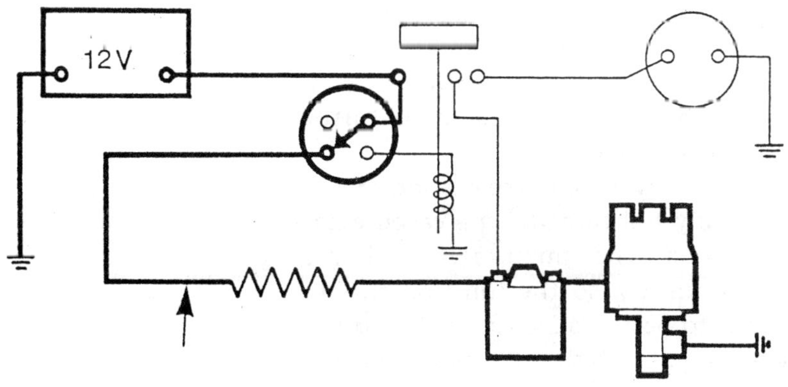

▲ *For normal running, current from the battery goes to the ignition switch and then through a ballast resistor, arrowed, to cut the voltage to the coil.*

▼ *With the ignition key in the 'start' position, the ballast resistor is by-passed so that the coil receives the full 12 volts. 1. is the starter solenoid, 2. is the starter motor, 3. is the coil and 4. is the distributor.*

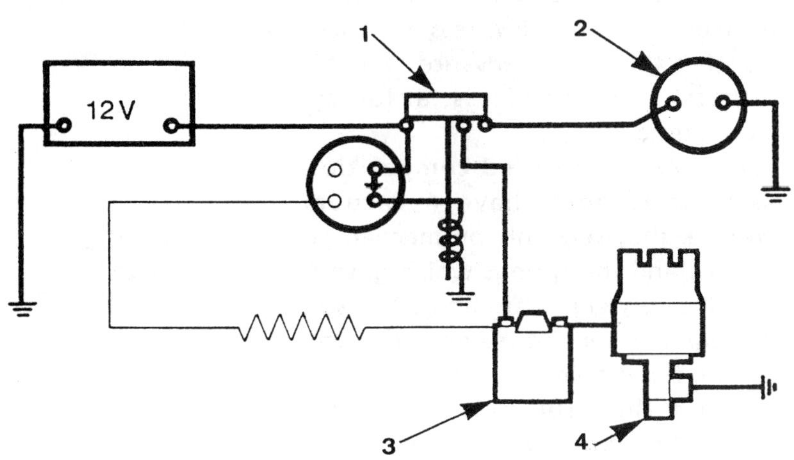

over-running the coil at 10 or 11 volts for the short period it takes to start the engine does it no harm. As soon as the engine fires and you let the ignition key return to the normal running position, the coil is fed through the resistance cable to run at its normal voltage. If it were left running at battery voltage it would overheat and its life would be considerably shortened, so it is important not to replace the resistance cable by a plain cable.

To understand uprating an ignition system it is as well to have a clear picture of what happens in the standard circuit. The coil is basically a voltage transformer with a low-tension primary winding which is energized by the battery, and a high-tension secondary winding which supplies the high-voltage current, anything from 31 to 40 kilovolts necessary to produce the spark across the plug gap. The increase in voltage is determined by the ratio of turns in the primary and secondary windings.

As with all transformers, a coil will produce a secondary voltage only with AC current or an interrupted DC current. Interrupted DC current is used in an ignition system, and the current is interrupted by the contact breaker points in the distributor. When the points are closed, current flows through the primary winding and energizes the secondary winding. When the points open, the current in the primary winding collapses very rapidly and the secondary winding discharges its voltage through the high tension lead to the centre of the distributor and from the distributor to the plugs. Timing the ignition is simply a case of arranging for the points to open at the exact moment that the plug needs its spark.

It is important for a good spark and accurate timing that the secondary high tension voltage in the coil discharges through the plug points as quickly and accurately as possible. The speed and accurate timing of this discharge depends on the current in the primary circuit collapsing as quickly as possible.

The enemy of a quick current collapse is a woolly instead of a sharp and accurate break in the primary circuit current at the distributor points. The break won't be sharp and accurate if they break with a big fluffy spark which they will do unless something is done to stop it. This sparking also burns the points quite rapidly.

The something that stops it is the capacitor, which used to be called the condenser and sometimes still is. The capacitor is a simple and cheap device which stores electrical energy but won't pass a direct current. It is connected across the distributor points, between the moving point and earth and, as the points start to open it provides a much easier path for the primary current instead of it's trying to jump the widening gap. It doesn't eliminate sparking at the points but it cuts it drastically. When the points close again the capacitor discharges its stored energy against the resistance of the circuit so it's ready for the next time the points open.

The reason for having an automatic advance mechanism in the distributor is that the compressed mixture in the combustion chamber takes a finite time to burn and expand irrespective of how fast the engine is running. It follows that, as the engine speed increases, we have to spark the mixture earlier – in other words advance the timing – so that it completes its burning and exerts the maximum downward force on the piston at the right time, just as it has passed over top dead centre.

Ford's distributor has both a mechanical advance, which advances the timing as the engine speed increases, and a suction advance, usually termed a vacuum advance even though the depression inside it never gets near a vacuum. This advances the ignition depending on the depression at the throttle of the carburettor, or the primary throttle in the case of a twin-choke carburettor.

The reason for having both is that, on an engine in standard tune, the ignition timing is varied both according to the engine speed and according to the load on the engine. The easiest way to monitor the load on the engine is the position of the throttle and, as the position of the throttle governs the depression in the carburettor choke barrel, using this is a convenient way of altering the ignition timing to suit the load on the engine.

When cruising on a light throttle, for example, the petrol and air mixture is quite lean, which gives good economy, but because a lean mixture burns more slowly it needs to be sparked earlier. At light throttle there is a good depression in the carburettor, so the vacuum advance unit on the distributor advances the ignition.

Conversely, when accelerating at full or nearly full throttle, the carburettor delivers a richer mixture which burns more quickly, so the ignition timing needs to be less advanced to avoid too early a spark which could cause pinking. As a wide throttle gives less depression in the carburettor, the vacuum advance unit advances the ignition less.

The reason for monitoring the depression in the carburettor choke barrel rather than in the inlet manifold is that there is a heavy depression in the manifold at tick-over on a closed throttle. This would give ignition advance when it isn't wanted, so the pipe from the vacuum unit on the distributor is taken to a connection close to the throttle butterfly on the carburettor where the depression at tickover is much less.

The two systems, mechanical and vacuum, work well together with the standard camshaft. However, when you fit a high performance camshaft with large valve overlap, you may run into problems with vacuum advance at tickover and low revs. The greater overlap of the performance camshaft tends to set up small pulse variations in the inlet manifold at low revs and these can affect the vacuum advance so that it 'dithers'. The result is a lumpy, uneven tickover.

This is why so-called competition distributors don't have a vacuum advance, they rely only on mechanical advance and the springs which govern the advance weights are altered to suit the degree of tune of the engine. If you run into this problem, but don't want to fork out the £100 plus for a new distributor, you can disconnect the vacuum advance and block it off at the carburettor to avoid sucking in air. You won't get quite the same degree of control over the advance curve, and possibly a little less economy at light throttle cruising revs because you'll still be using the standard advance weight springs, but at least you will get a smoother tickover and smoother running at low revs without detracting too much from the power output.

We will see how the two types of automatic advance work when we come to overhauling the distributor.

Basic ignition timing is carried out either with the engine static or running at tickover revs when the mechanical advance is inoperative and with the vacuum advance pipe disconnected and plugged so that the depression in the carburettor choke barrel at tickover doesn't affect the ignition timing. The timing is adjusted by rotating the distributor in its housing.

It is most important to accurately set the distributor points gap – the gap between the points when the fibre heel of the moving point is on one of the lobes of the distributor cam – before timing the ignition. If the gap is too small, the cam will open the points too early. If it is too large, the cam will open them too late.

It is also not generally appreciated that the points gap can affect the high-tension voltage at the plugs. We saw earlier that when the points are closed current flows through the primary winding of the coil and energizes the secondary winding. This has to happen every time a spark is required so that at high engine revs there are only a few milliseconds for this energy to build up in the coil.

Anything which cuts down on that time will affect the secondary voltage and the quality of the spark. If the points gap is set only a few thousandths of an inch too small the points open early and close late, and the time during which they remain closed to allow the high tension voltage to build up is dramatically cut.

For initial setting up of the engine, the gap can be set with feeler gauges to 0.64mm (0.025 in) but this is not accurate enough for high-efficiency high-speed running, so we use a dwell meter. Dwell, in a distributor, is the number of degrees of cam rotation during which the points are closed. We saw that altering the points gap affects the time during which the points are closed so, if we can measure that time we also measure the gap, and the easiest way to measure the time is by measuring the dwell angle.

Modern mass-production of electronic instruments has made home engine analysers, which measure dwell angle among other things, so cheap that it is foolish to try to set up a contact breaker ignition system for high performance without one.

Good, high quality high tension leads play an important part in engine tuning, particularly if you have fitted a high performance coil. You can buy cheap HT leads in many accessory shops and they work fine when they are new and dry. However, when they get older and, particularly when they get damp with condensation, they leak.

Sometimes this shows itself as a complete refusal to start on a damp morning but, more usually, the first signs are hesitation and a reluctance to accelerate brightly until the leads have thoroughly dried out. Sometimes on a dark, damp evening, you can see this leaking as a shimmer of sparks round the leads or, if you're brave enough, you can feel it as a high voltage jolt up your arm if you hold the leads. Top quality leads cost more but they don't leak like this till they get very old.

The law requires that your ignition system shall not cause radio or television interference, and plain copper-cored high tension leads will. They will also cause terrible interference on any radio you fit in the car.

This interference can be suppressed by fitting resistors in the high tension leads. You can buy wire wound resistors to fit in the leads, but they are not so effective, nor so efficient as modern resistor leads where the resistance is spread evenly along the cable's length.

Resistor leads are made from a centre core of graphite-impregnated carbon and are insulated on the cheaper ones by plain pvc. Medium-quality resistor leads are insulated with synthetic rubber and an outer sheath of Hypalon which will hold its insulation at temperatures up to 150°C. Top quality leads use synthetic rubber with an outer sheath of Vamac and will resist temperatures up to about 175°C. You can buy these cables by the metre and fit your own ends, but there is always a chance of water getting in at the connector. It is much better to buy them with the end connectors already moulded on.

For each type of insulation there are usually two ranges of cable resistance, 13,000 to 23,000 ohms per metre and 23,000 to 35,000 ohms per metre. With the standard coil, Ford specifies that the maximum resistance in any one lead shall not be more than 36,000 ohms, so if all your leads, including the longer one from the coil to the distributor, are under 1 metre long, use the higher resistance leads.

On a Crossflow tuned for top power, however, it could well pay you to experiment with different resistance leads and different plug gaps. You will probably be using a competition contact breaker distributor, or electronic ignition, because at engine speeds much above 5000 rpm there is a tendency for the contact breaker points on the standard Motorcraft distributor to bounce and cause misfiring.

In theory, with lower resistance leads, the energy from the coil is released more quickly, but the spark is not of such high intensity. With higher resistance leads, the energy release is slowed and the spark has a longer duration. Different electronic systems respond differently – so, if you change to electronic ignition, experiment with HT leads of different resistance and plug gaps between 0.025 and 0.045 in to get the best results.

The last components in the ignition chain don't have any working parts, carry on working far longer than than they should do and are all too often taken for granted – the sparking plugs.

The basic concept of a sparking plug hasn't changed since they were invented back in 1860 by Etienne Lenoir for his gas engine. They have, however, improved dramatically over the years in strength, electrical insulation, pressure sealing, resistance to chemical corrosion and resistance to spark erosion.

Plugs are classified according to their heat range, a system first evolved by Champion back in the 1920s. To operate properly a plug has to strike a compromise between the electrodes getting so hot that they cause pre-ignition and fouling up at low revs if the electrodes run so cool that they cannot burn off the deposits of combustion products. If these products build up on the insulation of the centre electrode they provide a path to earth across which the high tension current can leak and weaken the spark or, if they become severe, stop it firing at all.

Ideally, the temperature of the insulator shouldn't rise much above 850°C nor fall below its self-cleaning temperature which is around 350°C.

Plugs are usually described as cold, or sometimes hard, and hot, or sometimes soft. This sometimes leads to confusion as some people tend to think of a hot plug as being suitable for an engine with a high combustion temperature and vice-versa. In fact, a cold plug is one where the centre insulator is thick and stubby so it provides a strong heat path from the centre electrode to the body so the electrode runs cooler making it suitable for engines with a high combustion temperature.

A hot plug, on the other hand, has a slender tapering centre insulator so the electrode runs hotter making it suitable for engines with a lower combustion temperature which tend to foul up plugs more quickly. Modern good quality plugs resist both fouling and pre-ignition much better than plugs of even a few years ago. Generalizations are dangerous but, generally, if you increase the power of a Crossflow above a mild tune of, say, free-flow air filter and performance exhaust, you need to move up a grade in the heat range of your plugs.

Not enough attention is paid to the plug gap, yet all the tuning you have done on the engine can be wasted if the plugs produce only a weak spark. On a standard engine, stay with Ford's recommendation, but when you start to tune your engine for more power and change the heat range of the plugs, consult the plug manufacturer.

Typically, a plug's electrodes will erode away so that the gap increases by about 0.006 in in around 6000 miles. Performance will have started to fall off even though you may not have noticed it. Plugs should be cleaned and regapped every 6000 miles, more often if you tune for very high power. Many people check the plug gap with feeler gauges, but these can give a false impression because the earth electrode erodes only under the tip of the centre electrode and a feeler gauge bridges this depression. Gaps should be checked with a wire gauge, and you can buy these very cheaply in most accessory shops. Even though they may appear to be working well, most plugs fall off in spark efficiency after about 12,000 miles and should be renewed.

Many owners of tuned engines change from Ford's standard contact breaker ignition to an electronically controlled ignition but, for a standard engine or one with only mild tuning, the standard ignition system is quite adequate if it is maintained in top-line order. Even if you fit a proprietary optical trigger electronic ignition system, this may well use the standard distributor, and if this is badly worn you could be wasting your money.

Apart from the points, internal parts for Ford's Motorcraft distributor are becoming hard to find. You might find a Ford dealer with some left, but don't bank on it. The only resort if you're staying with the standard distributor and yours is worn is to hunt around for one in better condition or build one up with the best parts from several. You can pick them up quite cheaply in most breakers' yards.

Old distributor caps and rotor arms are a prolific source of ignition problems and weak sparks and, as they are available either as genuine Ford or pattern, it is worthwhile renewing them unless they are almost new.

With caps, the enemies are corrosion inside the turrets where the HT leads fit, erosion of the studs inside the cap, a chipped or sticking centre contact and cracking and tracking. Light corrosion inside the turrets and on the studs can be cleaned off with fine emery paper, but if the studs are badly burnt the cap needs renewing. A sticking centre contact can be freed or renewed.

Cracking, often just surface cracking, is self-explanatory and, when it gets carbon dust and other impurities in it, is quickly followed by tracking which is a burnt path either between the inside studs or the centre contact and the studs.

Tracking can also show as a burnt line where someone has carelessly scratched the surface of the cap either inside or out. The path has burnt because carbon or moisture has entered the crack or scratch and allowed the high tension current to leak across. Any sign of this means the cap is useless for a tuned engine, even though it might possibly allow a standard engine to run.

Rotor arms suffer from burning and erosion of the brass plate which distributes the spark to the cap studs and also, occasionally, from tracking and leaking. Light erosion can be cleaned off but any signs of tracking or leaking from the brass contact to the inside of the rotor where it fits over the distributor centre post will show as misfiring at high speed and, in severe cases, a refusal to run at all.

If you are staying with the standard Ford Motorcraft contact breaker distributor, the points, capacitor and rotor arm should be renewed as a matter of course. Wear points to watch for are rocking or sticking of the centre cam and plate, a punctured or sticking diaphragm in the vacuum advance unit, weak or distorted springs which control the mechanical advance weights and a centre shaft which rocks in the distributor body.

Complete Motorcraft distributors, or new parts to repair worn ones, are no longer listed by Ford though you may be lucky enough to find a few still in stock at a Ford dealer. You are more likely to be lucky with the advance weight springs than with other parts, but all is not lost because, apart from these springs and the spring which controls the vacuum advance unit, Motorcraft distributors for all capacities of Crossflow engine from 1100cc to 1600cc are the same, and parts are interchangeable.

A replacement Bosch or Lucas distributor costs upwards of £85, but second-hand Motorcraft distributors are plentiful in breakers' yards. If the budget is tight, then for standard running and mild tuning you may well be able to make up a good serviceable one from several.

The main reason for rejection is a centre shaft that wobbles in the distributor body. Ideally there should be no side shake at all when you try to move the skew drive gear at the end of the centre shaft, just as there shouldn't be when you try for side shake at the other end of the shaft inside the distributor body.

With a second-hand distributor there is bound to be a few thousandths of an inch side shake, but if there is a distinct wobble the distributor will never give you really first-class ignition because, as the shaft and the centre cam wobble, the points gap and the dwell angle, and thus the timing, will constantly alter.

If you replace the contact breaker points with an optical trigger kit from Lumenition, or similar, slight side shake which allows the optical trigger to move a few thousandths of an inch radially in the sensor shouldn't have any ill effect. With more serious wear you will be wasting your money fitting an optical trigger, as you will be likely to get misfiring at high revs.

Whether you use contact breaker points or an optical trigger, it is advisable to strip down the distributor, at least in part, to check it for wear and old sticky grease which could cause erratic action.

Inside the distributor you will see that

the baseplate assembly is in two parts, a lower plate fixed to the body and an upper plate which carries the contact breaker points and the capacitor. The upper plate can be rotated relative to the cam by an arm from the vacuum advance unit.

Before you start dismantling you can check the action of the vacuum advance and retard with a simple tool you can make quite easily from a bicycle pump. Reverse the washer inside the pump so that it sucks instead of pumps, cut the valve end off the pump connector and push it over the small pipe on the vacuum advance unit, the one where the pipe from the carburettor fits.

As you draw the pump handle gently back to create a suction, the vacuum unit should pull the baseplate round. If it doesn't, either the baseplate is sticking or the vacuum diaphragm is punctured. Pull the pump handle back gently. It doesn't take much suction to operate the unit and, if you jerk the handle back and create too much suction you could split the diaphragm.

If one is fitted, remove the radio suppressor from the side of the vacuum unit and take the circlip off the peg on the vacuum unit arm which pulls the baseplate round. Then, when you take out the two crosshead screws, the baseplate assembly lifts out.

The two parts of the baseplate can be separated after taking off the circlip from the post where the upper plate pivots. Be careful not to lose any of the washers or the light spring between the plates. Clean both plates and check that the top plate pivots easily on its post.

Two screws hold the vacuum advance unit on and, after taking this off, you can dismantle the mechanical advance mechanism. Next to come off is the plastic bump stop which stops the weights from flying out too far.

Before you unhook the advance weight springs, make a note of which one fits on which weight. One is either stronger than the other or it has an elongated hook at one end. The purpose is that one weight flies out at lower rpm than the other to give the initial advance, and further advance at higher rpm is given when the second weight comes into operation. When they were new, the springs were often colour coded, but this is usually worn off on old distributors, so mark each weight and spring with a dab of Tippex or a similar marker.

With the weights unhooked, you can take the felt wick from the top of the cam plate, take out the circlip and lift the cam plate off. The spindle on which it fits has a series of grooves round it to hold grease which will probably be hard and caked. Clean it of and repack the grooves with a lithium-based grease when you reassemble things.

If the shaft is a good fit in the body that's as far as you need to go with dismantling but, if you are making up a distributor from several you may want to take the shaft out to find one which is a better fit. The skew gear at the bottom comes off after you drive out a cross pin which is a tight fit. Support the gear when you drive it out or you can easily bend the shaft.

When you reassemble the shaft, coat the shaft with a turbine hydraulic oil such as Shell Turbo 41 or Mobil DTE Extra Heavy. When you put the skew gear back on you may find that the hole doesn't line up with the hole in the shaft. The hole is seldom drilled absolutely central and you've probably got the gear 180° out.

If you're making up a distributor from parts of old ones and you fit a different skew gear to the shaft you may not be able to get the hole to line up at all. In that case, put the two washers and the gear on the end of the shaft and turn the gear till its hole is at right angles to the old hole in the shaft. Put a 0.015 in (0.38mm) shim between the skew gear and the end washer and drill a new hole 0.125 in (3.20mm) diameter through the shaft using the hole in the gear as a guide. When you take the shim out, the shaft should have the correct end float. If you've got some shim plate, it helps to keep the shim in place while you're drilling if you cut it to a horseshoe pattern.

If you're going more than very mild with your tuning you will be spending a fair amount of money on a performance camshaft, gas-flowed head and so on, and you will get the best out of these if you also go for a new distributor. The firm from which you bought your other tuning gear will probably be able to offer you a performance distributor to go with it.

If you use a well-known performance camshaft like Piper or Kent Cams the same company can supply a new distributor, either a Bosch or a Lucas, with a high output coil to go with it. The coil will probably be 12 volt operation so you will have to replace the Ford ballast resistance cable with a plain cable. The advance curve will be tailored to match the characteristics of the camshaft you buy.

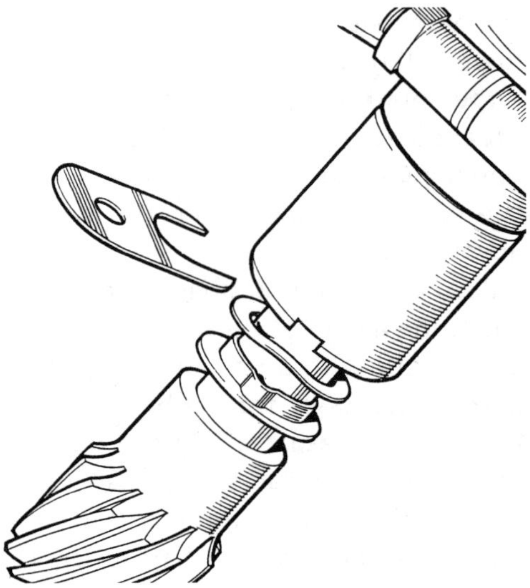

◀ *If you have to change the drive gear, a horseshoe shaped shim makes it easier to set the shaft end float.*

It may not have vacuum advance fitted, for the reasons mentioned earlier, but the advance springs and weights will have been chosen to give it the desired advance curve. If it hasn't got vacuum advance, remember to plug the vacuum take-off pipe on the carburettor or it will suck in air and upset the mixture.

It will still be a contact breaker distributor, and there is nothing wrong with a good contact breaker distributor for engine speeds up to 6500 rpm provided the points are checked regularly and the gap set with a dwell

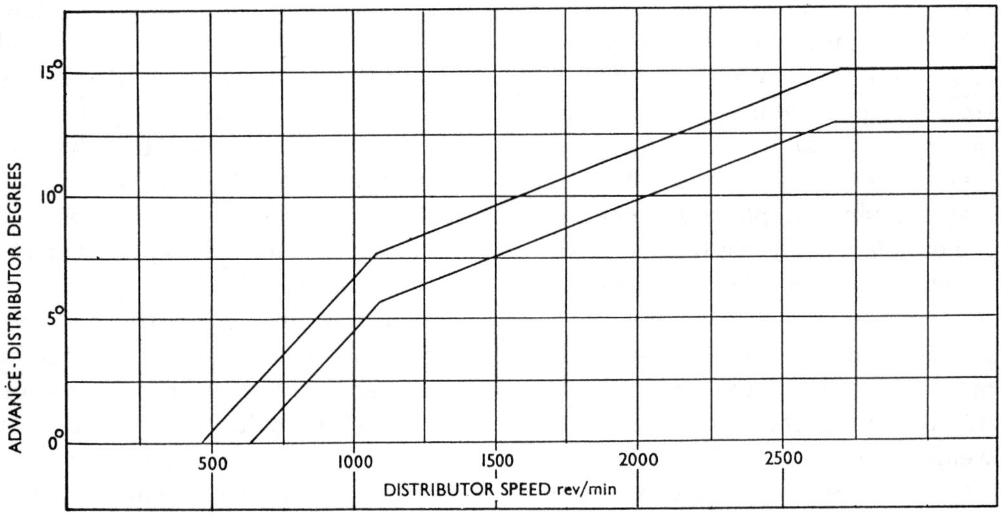

◀ *If you take your distributor along to a firm with an advance curve plotting machine, they will be able to give you a print-out like this for both mechanical and vacuum advance.*

▶ There is a hole in the baseplate through which you can use a screwdriver to bend the spring lug slightly to alter the mechanical advance curve. Bend the lug outwards to decrease the advance.

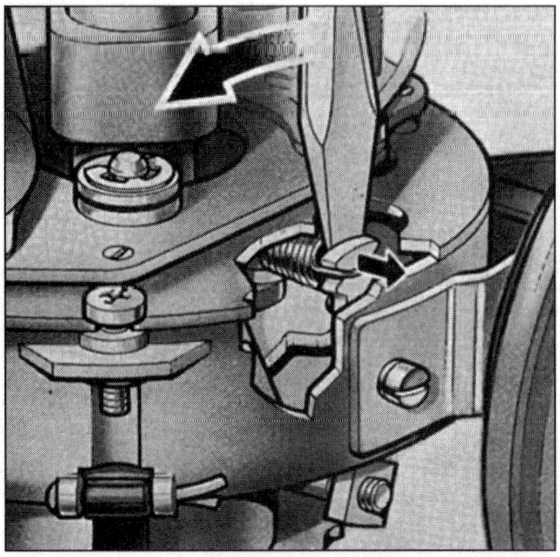

▶▶ In a similar way, you can increase the advance by bending the spring anchor lug inwards. Adjust the thinner spring first as this affects the advance curve at lower revs.

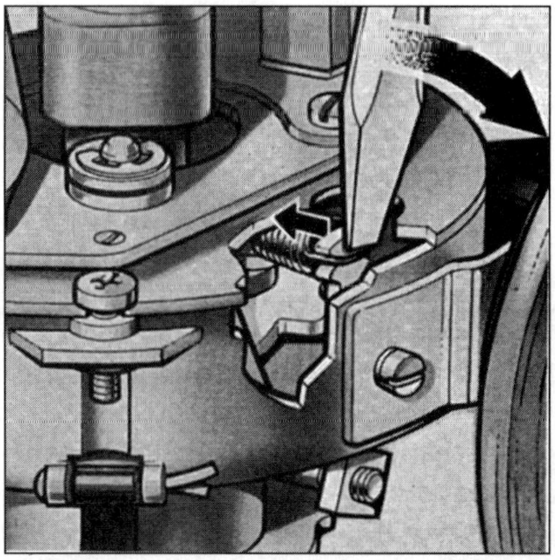

angle meter at least every 6000 miles. A new set of points should be fitted every 12,000 miles.

If you want to do less maintenance on the distributor, either a high performance one or the standard Ford Motorcraft, you can fit it with electronic control. There are two basic types. One still uses the distributor's contact breaker but this is now used as a trigger with a very low current so that the points last much longer. However, you will still have to check the dwell angle, and adjust the gap if necessary regularly because the gap closes as the fibre heel of the moving point beds down on the cam.

You may also find that, unlike a standard points system where, when the points burn, a pip develops on one point and a crater on the other, points which break with a very low current tend to develop a kind of surface skin, rather like oxidation. This has an adverse effect on the sharp break of the current so, for consistent results, the points still have to be cleaned regularly, even though they don't burn.

The second type, such as the Lumenition Optronic kit, does away with the points altogether and replaces them with an optical trigger. Once this is

▶ For initial starting, set the points gap to 0.025 in with a feeler gauge. Adjust it more accurately with a dwell meter when the engine is running.

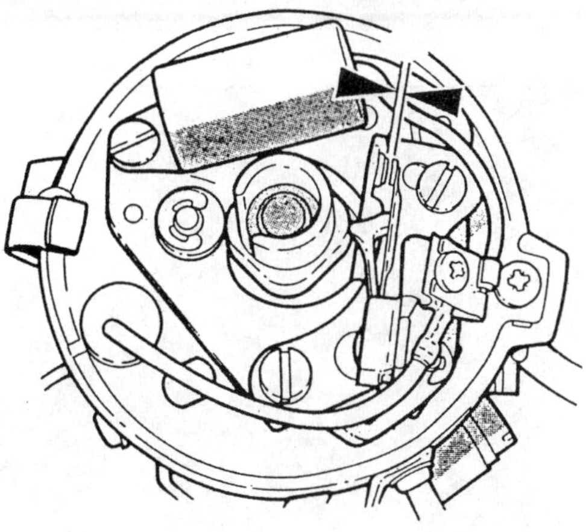

▶▶ In case you forget where the high tension leads go, this is a view from inside the distributor cap.

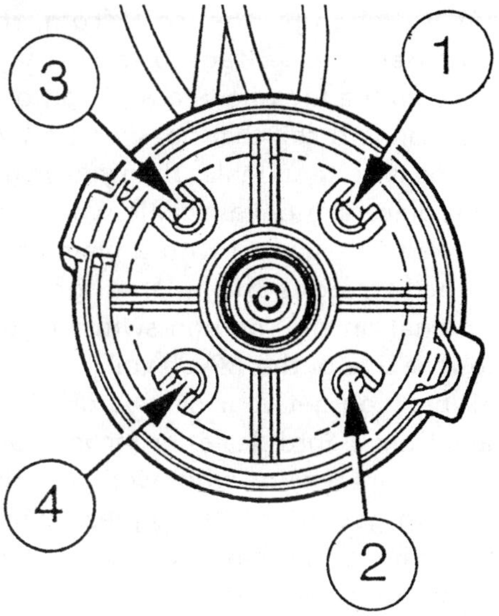

set up it will run for very long periods without attention and without falling off in efficiency. All that is needed is an occasional drop of oil on the mechanical advance and retard mechanism. For higher performance tuning, Lumenition also supply a complete performance ignition kit which, as well as the optical trigger in the distributor, gives you a high output coil and a higher performance electronic module.

Alternatively, you can get the benefits of electronic ignition by using the Bosch distributor from a Mk 4 Escort 1300 with the OHV Valencia engine (not the CVH overhead camshaft engine). The Valencia engine is derived from the older Crossflow and the distributor will fit straight in.

The mechanical and vacuum advance curves are designed for the 1300 Valencia, not the Crossflow, but I know of a Mk II 1600 Escort, tuned for fast road work, which has this modification and it certainly isn't lacking in power. As a final check, you could have a session on a rolling road where the advance curves can be plotted and, if necessary, they can be altered by changing the mechanical balance spring weights.

There are a few wiring alterations to do, but they're quite easy. First, you've got to change the coil from the ballasted low-voltage one normally fitted with a Crossflow to a 12 volt one. The standard coil that goes with a CVH engine is quite suitable. Then you've got to replace the ballast cable which fed the low voltage coil with a plain cable going straight from the ignition terminal on the ignition switch to the positive side of the 12-volt coil.

When you get your distributor, either new from a Ford dealer or second-hand from a breaker or wherever, you also need to get the multi-plug lead which goes with it. This has three cables in it – brown, green and black.

◀ *The distributor on the Crossflow normally has a top-exit cap.*

▼ *If you fit a special manifold, things may be tight and you will need a side-exit distributor cap.*

◀ *If you want electronic ignition you can use the Bosch distributor normally fitted to the Mk 4 Escort.*

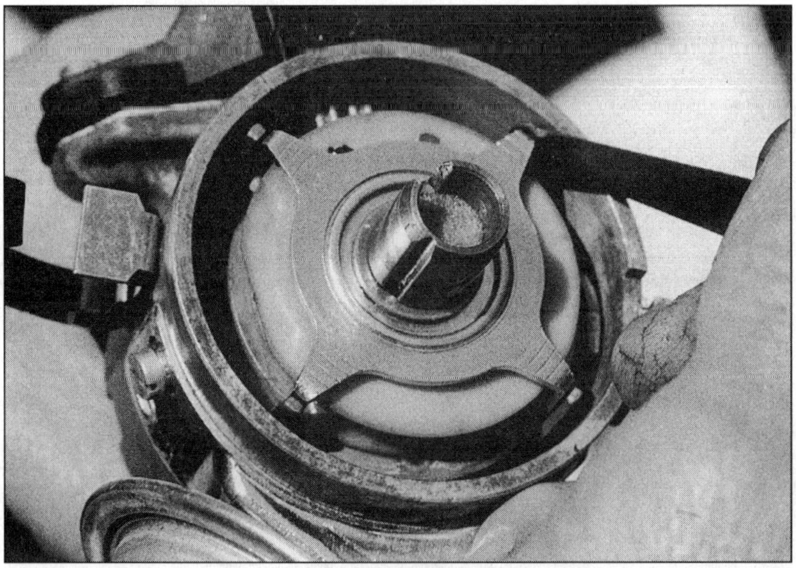

▲ *On a second-hand electronic distributor, check that no-one has bent the rotor trigger arms or stator posts. The gaps should be even at about 0.012 in, but the figure isn't critical.*

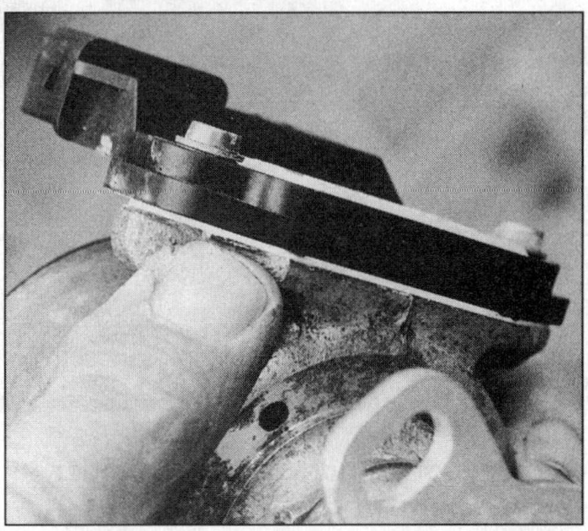

▶ *The amplifier module on the Bosch distributor must make good heat contact with the body. If necessary, clean the mating faces and use special heat-conducting paste.*

The green cable goes to the positive side of the coil, the terminal where you fitted your 12-volt feed from the ignition switch. The black cable goes to the other side of the coil (the negative side) – the side which also takes the cable to the rev counter if you've got one fitted. The last cable, the brown one, goes to chassis earth.

If you buy your distributor second-hand there are a couple of checks you should make on it as well as the usual checks to the cap and rotor arm. Most problems with the amplifier module on the side of the body come from overheating, and the main way the amplifier gets rid of its heat is to the body of the distributor.

In time, the contact between the two becomes poor, so take the amplifier off and clean both mating surfaces thoroughly. It doesn't hurt to bring the area on the body of the distributor up bright with fine emery paper. Then coat the surfaces liberally with the special heat conducting compound – which you can get from Ford dealers – before you screw the amplifier back on.

The other check is to make sure that the arms of the trigger underneath the

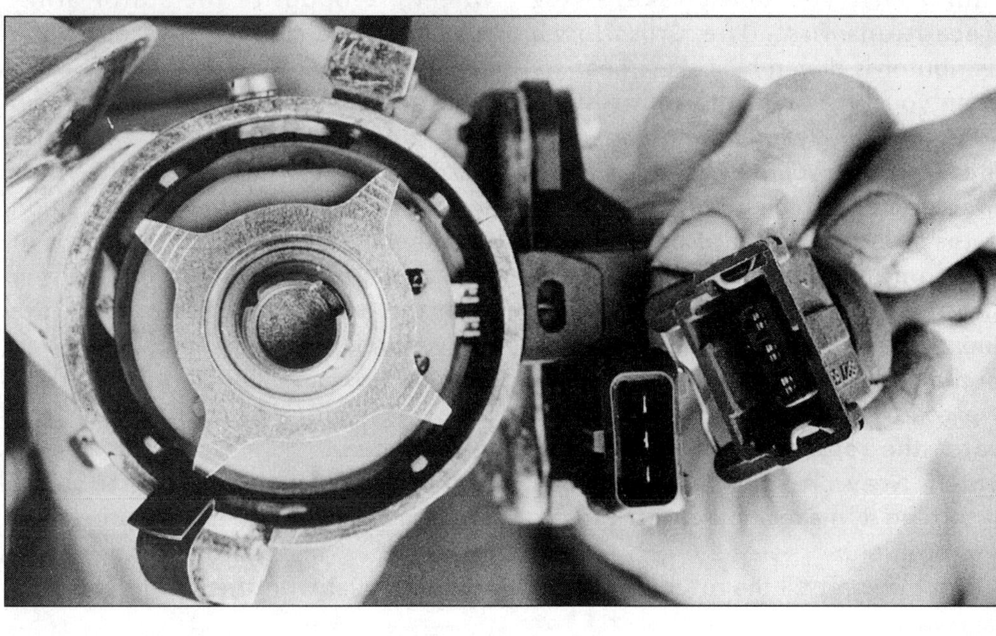

▶ *With the Bosch electronic distributor you will also need the wiring loom.*

dust cover inside the body all clear the stator arms properly. You shouldn't find any problems here, but some ham-fisted people have been known to bend the trigger arms. The gaps should all be the same, give or take a few thousandths of an inch, and a gap of about 0.012 in (0.30mm) is about right, though it's not critical. You'll also have to fit the fixing clamp from your old Motorcraft distributor.

◀ *When you by-pass the ballast resistor, the connection from the ignition switch goes to the positive side of the coil (A). The negative side (B) goes to the distributor.*

As standard on the Mk 4, the Bosch distributor is fitted with a top-exit cap. You might find that this is too tall when you try to fit it in your engine, particularly if you've fitted a different inlet manifold. You can buy a side-exit cap from Bosch agents, but it's rather expensive. Ford doesn't list one for the Mk 4.

However, turn to the Ford Motorcraft all-makes spares catalogue and you'll find a much cheaper side-exit cap in there that fits perfectly. It isn't listed for a Ford, it's listed for the Bosch distributor on a Fiat Strada Super 75 and 85. The Motorcraft part number is 5005566.

▲ *For a side-exit cap for the Bosch distributor, look in the Motorcraft catalogue. It's listed for a Fiat Strada.*

When you fit your new or rebuilt distributor you'll have a fine old game timing it if you don't take a few precautions first. The Crossflow's distributor is driven by a skew gear so, when you lift it out, the centre shaft turns. Similarly it turns in the opposite direction when you put it back.

To get a basic timing, turn the engine to the firing point with No. 1 cylinder on compression, and feed the distributor in. Don't worry about timing it for the moment. Make a mark on the body in line with the brass tip of the rotor arm. Now lift it out and watch the rotor arm turn. Remember which way it turned. Without disturbing it, make another mark for its new position.

Turn the shaft till the rotor is pointing at the stud for number one cylinder in the cap and the arms of the trigger wheel are opposite the stator arms. Now turn the rotor arm through the same number of degrees and in the same direction as the rotor turned when you took it out. As you feed the distributor into the engine, the rotor should turn back to the firing position for number one cylinder. If the arms of the trigger wheel aren't quite lined up with the arms on the stator, turn the body slightly to line them up before you tighten the clamp. Check again that the rotor is pointing to number one cylinder stud.

With a contact breaker distributor, you follow exactly the same procedure except that, instead of lining up the trigger wheel with the stator, you turn the shaft till the rotor is pointing at the stud in the cap for No. 1 cylinder and the points are just about to open.

This timing will give you a setting near enough to start the engine and you can do the finer timing with a strobe lamp in the usual way. As a starting point, try timing it with an advance of about 8° or 9° before TDC – vacuum disconnected and plugged, though. Depending on what camshaft you've fitted, you may find that anything between 6° and 12° before TDC suits your engine better. Try to get the advance curve checked on a rolling road if you can.

If you haven't got a strobe lamp you'll have to resort to the old play-it-by-ear method. With the engine ticking over at operating temperature and the vacuum connected, turn the distributor body slightly back and forth till you get the fastest tickover speed. Slow the engine down on the slow-running throttle screw and turn the distributor back and forth slightly again to see if the speed alters. Stop at the fastest tickover. Now try it on the road, and adjust it slightly if necessary. Keep an eye on your engine temperature just in case you're a little too far retarded. Not very scientific, but it works, and it will be as near as you'll get without a strobe.

▼ *This kit from Lumenition converts a Motorcraft contact breaker distributor to optical trigger electronic ignition.*

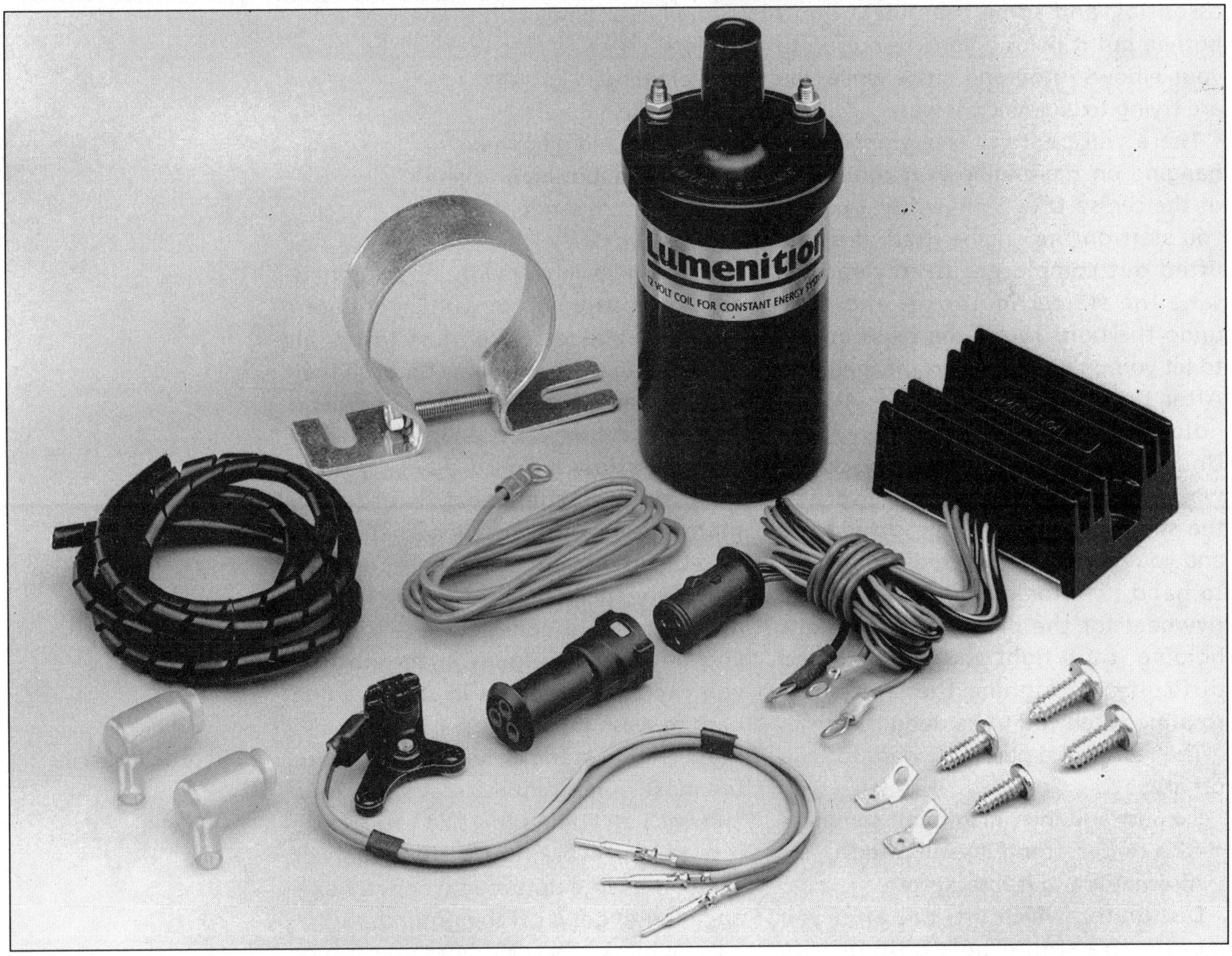

Chapter 7

Stripping the engine

The first stage of stripping your engine, after making sure that the oil has been drained out, is to give the outside a good clean off with paraffin or white spirit. It isn't absolutely essential, and some rebuilders don't bother, but it helps if you are not up to your elbows in oil and muck while you are trying to examine the parts.

There will be quite a few ancillaries hanging on the engine as it comes out of the car, so take these off first before you start on the engine itself. If it was lifted out complete with its gearbox, take the starter motor off and then undo the bolts round the bell housing to let you pull the gearbox off the back. After that you can undo the six bolts holding the clutch to the flywheel. Undo them a turn or two at a time, working diagonally, or the pressure of the springs may distort the cover plate and you may find it rejected if you come to hand it in for exchange. Leave the flywheel for the moment as the bolts holding it are tight and you may find difficulty in stopping the crankshaft rotating when you try to undo them.

The rest of the parts you want to take off are:

Exhaust and inlet manifolds, complete with carburettor if it's still fitted.
Alternator and its bracketry.
Distributor, which lifts out after you undo the single bolt holding the clamp plate.
Sparking plugs.
Water pump, thermostat housing and thermostat.
Crankcase emission oil separator, which fits just behind the fuel pump and has a steady bolt to the block.
Fuel pump.
Dipstick.
Oil filter and oil pump.
Engine mounting brackets.
Oil pressure sender unit.

This will leave you with a bare engine ready for stripping down. Experienced engine rebuilders often just throw all the bolts together in a tray because they know from experience which fit where. Unless you're experienced in working on the Crossflow I would advise you to keep them in some sort of order. In many cases you can screw them back into their holes but, if this isn't practical, get a few small tins, food tins are useful, and keep sets of bolts in different tins. If you punch a few holes in the bottom you can shake the tins in the cleaning bath to wash the bolts.

An engine stand is nice to have, but for most home rebuilders it's a luxury, and with an engine this size you can do just as well working on the bench or even on a large upturned wooden box – anything to get it off the ground.

Put a block of wood under one end of the sump to stop the engine rocking, and start at the top end by taking off the rocker cover. You can get a reasonable idea of what you're likely to find deeper down by the state of the engine under the rocker cover. There is bound to be a lot of black oil around, but this is usual.

What is less reassuring to find is a substance that looks like thick black butter. This means that the engine oil hasn't been changed for thousands of miles and is usually a sign of worn bearings and worn bores.

Undo the bolts holding the rocker shaft evenly and gradually, as some of the valve springs will be under compression and make the shaft tilt as you loosen it. At this stage, give the rockers a glance to see how badly the pads are worn but don't delve too deeply for the moment. Put the whole assembly in a wire basket, or a tray, to be cleaned off.

For a complete engine recondition, when you're looking for a big upgrade in power, you will be using new cam followers, and probably new rockers, so there isn't a lot of point in keeping the push rods in order. If, on the other hand, you're on a strict budget and you just want to give your engine a new lease of life with just a mild pep-up, you'll want to use as many parts as possible again.

Parts which have bedded down together don't like mating down with other part-worn bedfellows, so keep the push rods in order. One way is to get a length of plywood, or even a length of stiff cardboard, make eight holes in it, mark one end 'FRONT' and push each push rod in its appropriate hole as you take it out. Alternatively, we have known some people who get a length of string, make two knots in one end to indicate the front and then tie the string round each push rod as they take it out. The advantage of this method is that you can drop the whole lot into the cleaning bath without losing the order.

Next to come off is the cylinder head and, strictly speaking, you should just 'crack' the bolts half a turn or so – in the reverse order in which they are torqued – before you fully undo any. This lessens the chance of the head warping. If you can't remember the order, work in an 'X' pattern from the ends inwards for the outer eight bolts and finally the two in the middle of the head.

Sometimes the head will come free as soon as you undo the bolts, but if anyone has used gasket cement on the head gasket – which they should not have done – or sometimes even if they haven't, the head can stick surprisingly tightly. Don't try hammering a screwdriver in to lever it up, put one centre bolt back loosely to stop the head sliding off the block and use either a heavy mallet or a hammer and a block of wood to knock the ends of the head sideways to free it. You'll be glad you remembered to take the sparking plugs out because, if you forgot, they might catch on the edges of the cylinder bores when the head swings round.

In its manual, Ford specifies a valve spring compressor that bolts on the top of the head, but a normal G-clamp type of compressor is just as effective – easier, in my opinion. There's a little tip about compressing valve springs that you may not have come across. In use, the valve retaining collets jam very hard in the tops of the spring retainers so that it's sometimes difficult to get the initial movement on the compressor. Before you start, go along the spring retainer caps with a piece of metal tube or a socket and a hammer. Give each one a sharp blow downwards to free the collets. They won't jump out, but watch your fingers as the tube or socket often bounces back quite smartly.

If you intend fitting new valves don't

worry about keeping the old ones in order, but if you do intend using them again, make eight holes in a strip of wood, label one end 'FRONT' and pop each valve in its appropriate hole as you take it out. Don't bother to keep the old valve springs, they'll be past their useful life and so will the valve oil seals. If you're buying a performance camshaft kit, it may well include new spring retainer plates – it certainly will if you are going over to double valve springs, but check before you throw the old ones away. Keep the valve collets as you don't normally get new ones with an overhaul or tuning kit.

It's now time to turn the engine upside-down (with the sump uppermost). Although you drained out the old oil, there will still be a surprising amount left inside, so spread a couple of layers of old rag on the bench to catch it.

Taking the sump off is straightforward enough, except that it usually sticks tight to the gasket. This is where you have to use a lever, and I recommend a very useful tool usually called a pry bar. It's shaped like a large screwdriver with the blade end turned up at a slight angle. You'll find all sorts of uses for it in engine dismantling – and, come to that, for other jobs on the car.

Inside the bottom end, the first thing to take off is the oil pick-up pipe and filter, because it sticks up and gets in the way. The stem of the pick-up pipe is a tight push fit in the block and there is a single steady bolt holding a flat bracket to the crankcase. The pipe might pull out of the crankcase quite easily, or it might be very tight. There's nothing else holding it, so, if it's tight, persevere with lifting and twisting until it comes out.

Now you can take the flywheel off because you can use a block of wood to stop the crankshaft turning while you undo the bolts. Use the block of wood as a wedge between the rearmost web of the crankshaft and the side of the crankcase. After you take the flywheel off, put a couple of the bolts back in the end of the crankshaft, they're handy to put a bar between to turn the crank over while you're undoing the big-ends.

You can also use your block of wood at the front end to stop the crank turning while you undo the bolt holding the fan pulley. You may find both a spring washer and a flat washer under this bolt. The pulley might come off relatively easily. If not, use a three-legged puller or, if you haven't got one, a pair of tyre levers. It usually comes off after jerking it a couple of times. Get the levers down behind the pulley and lever as near to the centre boss as you can to avoid distorting the pulley flange.

When you take the timing cover off, note that the concave oil thrower fits with its concave side outwards, towards the front. Undo the snail cam of the chain tensioner and lift the tensioner blade off its pivot post on the block. The fibre pad will undoubtedly be well worn, but don't throw it away until you've got a new one.

After you've taken out the camshaft sprocket holding bolts, locked with locking tabs, you can use your pair of tyre levers to prize both chain wheels off complete with the chain. Even on a standard engine rebuild it isn't advisable to use the old chain again. If you're going to fit a vernier timing camshaft chainwheel you won't want the old one and if you're looking for lots of power increase you'll be fitting a duplex chain and a new pair of wheels, so you won't want the old crankshaft chainwheel either, but don't throw anything away until you've got the new ones.

Put the chainwheel bolts back into the end of the camshaft and use a bar to turn it a few times so that all the mushroom headed cam followers are lifted. I say 'lifted' because that's the natural way to think of it, but actually they will be moved downwards free of the camshaft because the engine is upside-down. If any stick, use a thin drift and a hammer to push them down. Pull the camshaft out from the front of the block, but as you do so be careful that you don't catch any of the sharp-edged cams on the bearings. On a very high-power rebuild it may be advisable to fit new camshaft bearings, but for a straightforward rebuild, or even a fast-road engine, the camshaft bearings can usually be used again because they seldom wear very much.

With the camshaft out, you can lift out the cam followers. An easy way to do this is to use a magnet on the end of a rod but, if you haven't got one, you can lay the block on its side and push the cam followers out with a rod from the top of the block. It's usual on a rebuild to fit new cam followers but, for the moment, keep them in order. There were two types of cam follower used on the Crossflow, one with a larger stem diameter than the other. The larger one is fitted on 711M engines. You will need to specify which you want when you buy the new ones.

Use a bar through two bolts in the flywheel end of the crankshaft to turn it so that you can undo the big-end bolts. Notice that each rod and cap is numbered on mating faces at the side, with No. 1 at the front of the engine. After each pair of bolts is undone a few turns, give the cap a light tap with a hammer to free it, then take the bolts out and lift it off. The pistons and rods come out through the top end of the block. Throw the old bearing shells away, but bolt each cap to its respective rod to avoid mixing them up.

The last job before lifting out the crankshaft is to remove the oil seal housing and seal from the back. Then you can undo the main bearing cap bolts, lift them off and lift the crankshaft out. Note that the bearing caps are numbered but, unlike the big-end caps, they are not numbered from 1 to 4. The front cap is marked 'F', the centre one 'C' and the other two are marked '2' and '4', usually with a letter as well as a number. Note that there are thrust washers in the crankcase either side of the centre main bearing.

The engine is now completely in pieces but there's one last job before you start cleaning the components, driving out the old water jacket core plugs. Ford uses quite large core plugs, shaped rather like the lid of a shoe polish tin, with the hollow side outwards. To remove them, use a punch or an old screwdriver to hammer one side inwards. The plug will tilt over in its housing and you can either lever it out or pull it out with a pair of pipe pliers.

Turn the block the right way up again and dig down from the top in the water jacket with a long screwdriver to get as much sediment out through the core plug holes as you can. You'll find that there is usually more at the back where the water flow is least.

It makes the job of examining and checking all the parts much easier and more pleasant if they are all cleaned off first. You can clean off many of the components in a cleaning tank at home using a proprietary degreaser such as Gunk or Jizer or, as I usually do, in a bath of ordinary white spirit which you can buy in large plastic cans from a paint materials supplier or a DIY supermarket.

With the parts all cleaned and dried off you can go through them deciding which you can use again and which you have to replace. I find it useful to get all the 'non-working' parts checked and attended to if necessary before going on to the parts which wear. It makes rebuilding the engine much more pleasant if everything you pick up is ready to fit and you don't have to keep stopping to fettle something which is damaged or distorted, or, even more frustrating, have to stop work to go and buy a part which is damaged beyond re-use. While you are checking, make a note of any bolts, screws or washers which need replacing, and get a stock of new ones before you start the rebuild.

Unless you are going to use a fancy alloy rocker cover for looks, or unless you are going racing and want a fully baffled sump or possibly a cast finned one, you will want to use the rocker cover and sump again.

Most rocker covers on old Crossflows leak at the joint face, and this is almost always because successive over-tightening of the fixing screws to try to stop the leak has raised domes and burrs round the bolt holes. Turn the cover upside down, find a block of metal which will fit under the flange and go along with a light hammer dressing it down to make sure that it is flat.

Finish off by running a file lightly along the flange to highlight uneven patches, and dress it again until you get it as flat as you can. The new gasket will take care of small irregularities but not distortion. It's worthwhile spending time doing this properly as it can be most annoying on a newly rebuilt and gleaming engine to find oil running down from joint faces that ought to be oil tight.

Do the same with the sump and, if the bottom has been dented by someone dragging it along the floor, beat it out with a soft-faced hammer or a block of wood and an ordinary hammer. Check the thread in the sump drain hole and the boss for flatness, and check that the drain plug hasn't been burred and rounded off by pipe grips. If it has, get a new one and a new sealing ring. You don't often get a drain plug sealing ring in an engine overhaul gasket set. Check also that the curved recesses which hold the oil seals at the front and back aren't bent or distorted.

There's a point worth watching about sumps. If you are rebuilding a 1600 Crossflow and have to look for another sump for any reason, make sure you get one from a 1600 engine. The sump from a 1300 looks the same and will fit on, but the pan is slightly narrower so that the webs of the 1600 crankshaft, which has a longer throw, just rub on the sides of the pan. It's so slight that you probably won't notice it when you turn the engine over by hand during assembly but, when you start up, it creates an awful clatter and you wonder what on earth has gone wrong.

It's not very likely that the joint face of the front timing cover will be distorted, but check it with a straight edge to make sure as this can be another annoying place for an oil leak. On Crossflow Fiestas, with their thinner timing covers, you sometimes find that a slack chain has rubbed against the inside and worn a groove. In bad cases it can even rub right through. Check the housing where the crankshaft oil seal fits, to make sure no-one has damaged it by digging out the old seal with a screwdriver.

It is false economy to refit the old clutch unless it was renewed a short time before the engine was stripped, but you will want to save the flywheel if you can. The only reason for rejecting a flywheel is if the clutch face has become so ridged that it cannot be refaced. Light ridging and 'Chinese writing' surface cracking can be taken out by having the clutch face reground, either by a local engineering firm or by the specialist tuning firm you are using to work on the rest of the engine.

The teeth of the flywheel ring gear where the starter motor engages will probably be worn, but the gear is replaceable. It Isn't a Job you can do at home as it is a shrink fit. After the old ring gear is cut off, the new gear has to be heated to expand it before it is put on and, if it is heated unevenly, it will distort.

There are two types of ring gear – one for inertia starters and one for pre-engaged starters. The one for inertia starters has chamfers on leading edges of the teeth, and the one for pre-engaged starters doesn't. Make sure you have the correct new ring gear fitted. In the case of a ring gear for an inertia starter, some have a chamfer both sides, but some have it only on one side. Check when you get the flywheel back that – if it is a single sided chamfer – the gear has been fitted the right way round.

Pay particular attention to cleaning out the oil pick-up pipe and its strainer, and check the pipe for distortion. Make sure that the part which pushes into the crankcase is cleaned of old jointing compound or you will have problems replacing it. Before push-fitting it back give the end a smear of Loctite to seal it.

Early 1100 Crossflows were fitted with a simple draught-tube crankcase breather but this was soon replaced by a semi-closed ventilation system. Ventilation starts by air going in at the oil filler cap through a breather valve. The cap should have been cleaned out every 18,000 miles, but it probably wasn't, so either clean it or renew it.

Crankcase oil fumes flow out through an oil separator, a small metal box with internal baffles which fits into the boss which takes the mechanical fuel pump. An emission control valve sits in a grommet on top of the oil separator and, from there, a pipe goes to the inlet manifold just below the carburettor. On a few models, the emission valve was fitted at the base of the carburettor Instead of on top of the oil separator.

The emission valve can be taken apart for cleaning after taking out an internal circlip. Wash and dry all the parts before putting it together again, and wash out the oil separator. The grommet is an oil resistant material and may still be in good condition, but it may need renewing. Following the general principle of never throwing anything away until you've got a new one, keep it just in case. Renew the pipe from the valve to the inlet manifold.

While you're at this end of the engine, check the dipstick tube – which is a press fit in the block. They are sometimes bent by careless fixing of hoisting chains when the engine is taken out. If you are rebuilding a second engine while yours is still in the car, check that you have a dipstick and that it is the correct one for the engine. It sounds elementary, but different models of Crossflow had different sumps and different dipsticks.

If you can, it's best to take the block and cylinder head to someone with a hot cleaning tank which will get the old oil deposits out of all the recesses much more easily than you can with a brush. Most firms which rebuild engines will do this cleaning for you at a small charge. Before you send the block to be cleaned, take the end plugs out of the oil galleries. They're screwed plugs which take an Allen key.

When you get the block and head back they will look very clean but, even so, a good maxim when tuning an engine is never to take on trust anything done by someone else. Dirt is the enemy of every engine, so use a syringe – a plastic garden syringe is quite suitable – and squirt paraffin through all the oil passages in the head and block until you are certain they are free and clean. Remember to replace the plugs in the ends of the oil gallery.

If you are cleaning the crankcase and head at home, don't be tempted to rush it. Stand them in a tray and clean with a brush and solvent until you're sure that there are no pockets of old oil deposits left; then syringe out all the oil passages.

Before you go on to examine the crankcase and head for running wear, carefully go over with a straight edge all the smaller joint faces – timing cover, oil pump, fuel pump, sump and so on – to check that they are true. It's most unlikely that they will be warped, but someone who had previously worked on the engine may have used a screwdriver or lever to take components off and may have damaged the face. If necessary, dress the joint face with a large, fine-toothed, flat file.

All the threaded holes where components bolt on should have light counterbores so that the thread doesn't pull up proud of the joint face and stop the gasket seating properly. You are unlikely to find any with a proud thread, but it has happened, so if you do, put a light chamfer on the edge of the hole with a drill several sizes too large. You don't need a power drill to do this – it's easier and safer to put the drill in a carpenter's wood brace. With a sharp drill, only two or three turns are necessary.

All this may sound like a counsel of perfection, but they are points which can give problems later on, and you will not find them checked on cheap reconditioned engines where speed is the essence of profit.

Next, we come to the major joint faces, the top of the block and the underside of the head. Here, you need a straight edge long enough to reach from the front to the back, and feeler gauges to measure any clearance under the straight edge where the head or the block face has warped.

I am not going to lay down how much clearance is too much. It depends on how far you are going in your power tuning. For top power on a racing or rally engine, any clearance is too much, and having the head and block refaced is almost mandatory. I would also advise it if you are thinking of going for higher fast road power later on, even though you may be starting off with a mild tune. The same applies if you are buying an exchange tuned cylinder head, which will have been refaced. Even though it is almost always the head and not the block which warps, it is silly spoiling the ship for a ha'p'orth of tar.

The decision becomes more difficult if you are just overhauling the engine without reboring or new pistons. I have overhauled engines with up to .003 or .004 in warping between the head and block without having any long-term problems. As a general counsel, if you are having the block rebored, have the block and head checked and, if necessary, refaced as well.

Much the same applies to cylinder bores. These don't wear evenly – the maximum wear is usually just under a third of the way down from the top. If you can feel any ridge at the top, however slight, then if you are tuning the engine at all, have the block rebored with new pistons.

If you are just overhauling and there is only a suspicion of a ridge, not even enough to catch your fingernail on when you have cleaned the carbon off the top of the bore, you can get a new lease of life by cleaning the pistons and fitting special oil control rings such as Cords. I must stress, however, that these are designed to extend the life of an engine in standard tune. If you try to use them as a substitute for reboring on a worn engine, or if you fit them in place of a rebore on a tuned engine, you are heading for disappointment.

If you use special oil control rings, make sure that the top ring is a 'ridge dodger' type, sometimes abbreviated to RD. This has a small shoulder cut out at the top face so that it doesn't clatter against the ridge. Make sure when you fit it that you fit it the right way up.

Before you have the block rebored, talk with the firm which does it about the pistons you intend to fit and get them before the reboring starts. It's important that the cylinders are bored and honed to fit individual pistons.

▶ *Undo the rocker shaft bolts evenly because some of the valve springs will be under compression.*

▶ *If you're using the rockers and cam followers again, keep the push rods in order. If not, it doesn't matter.*

◀ *Things will be pretty mucky when you take them off, so get them off the bench and into a cleaning tank as soon as possible.*

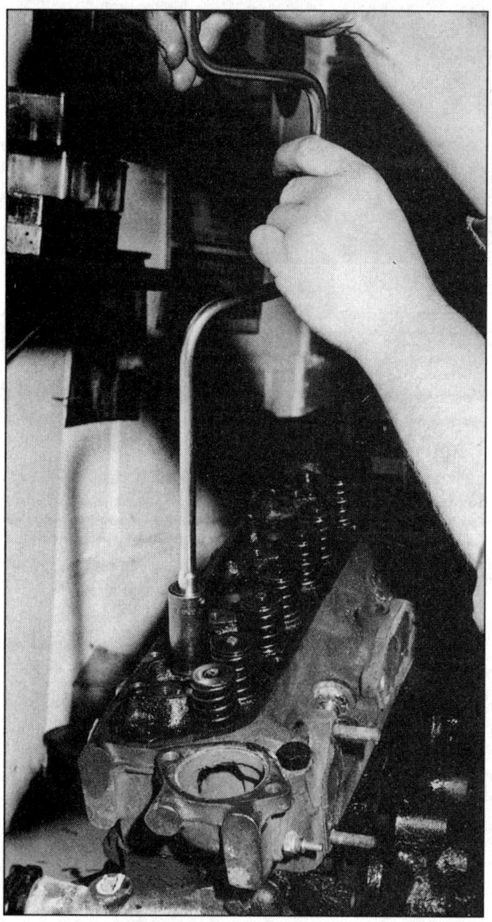

◀◀ *After 'cracking' the cylinder head bolts loose with a socket bar you can spin them off with a speed brace.*

◀ *The inside of an engine can tell you quite a lot. This one was quite good, with no signs of piston burning, very little carbon and no gasket leaks.*

▶ *The head gasket on this engine had been stuck down with compound, an unnecessary (and wrong) thing to do if the faces are flat.*

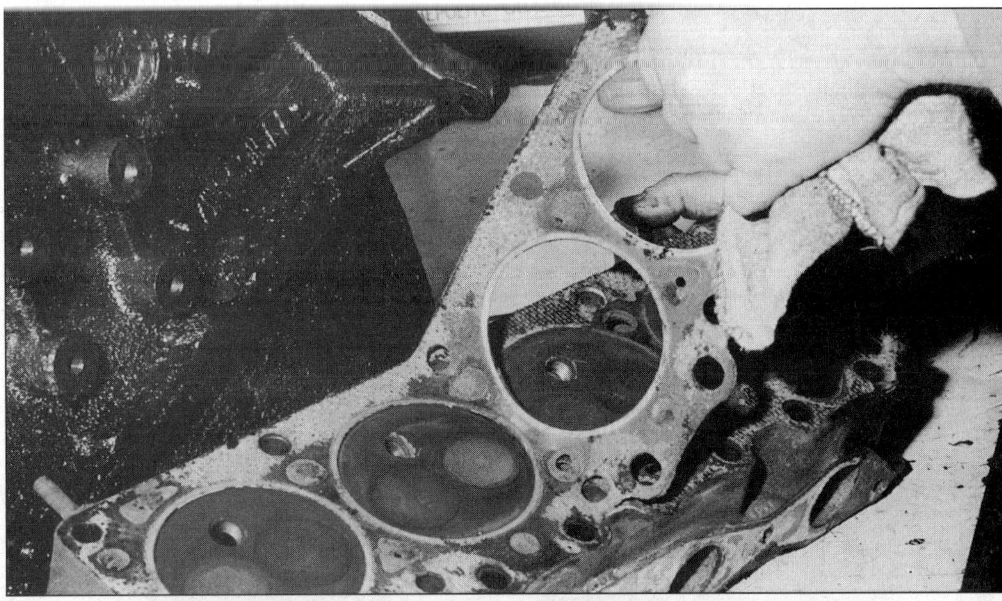

▶ *All the valve heads here look in good order and the carbon deposits are light and soft – a sign of detergents in the fuel.*

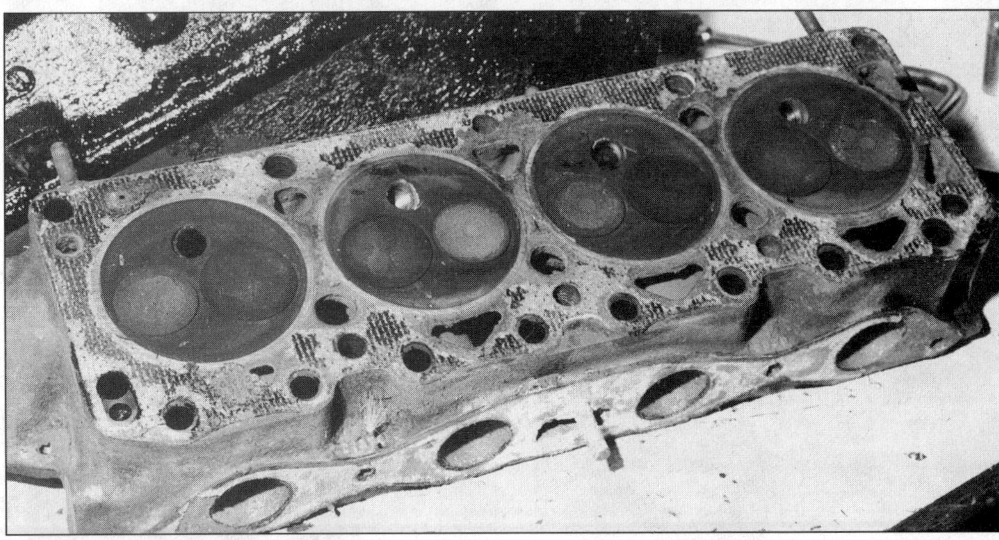

▶ *Next to come off the front of the engine is the water pump.*

▶▶ *This water pump looks OK, but they don't give warning signs when they fail, they just leak, so renew it as a matter of course on a rebuild.*

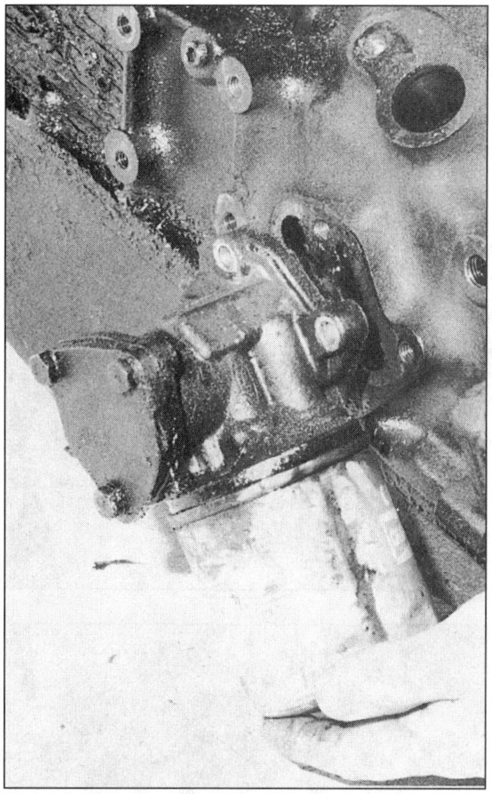

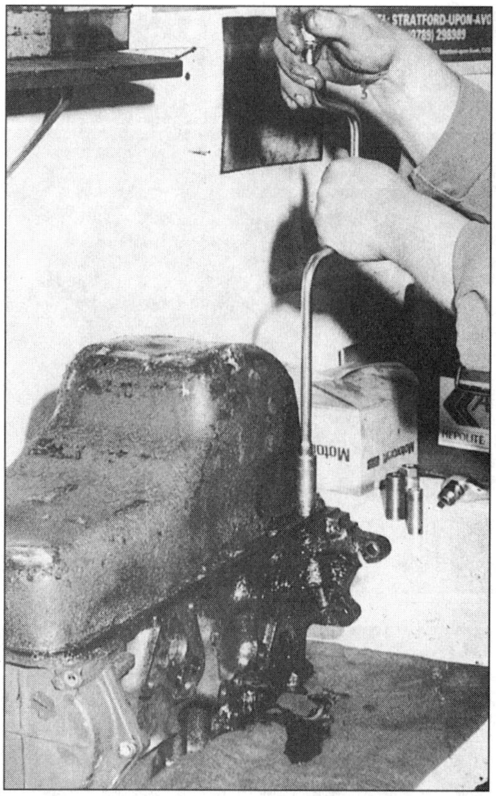

◀◀ *If the oil pump is in good condition when you strip and examine it, you can put in an uprated relief spring and plunger and use it again.*

◀ *Now the engine can be turned upside-down and the sump taken off.*

◀ *Sumps very often stick, and this is one place where the pry bar comes in handy, but try not to distort the sump edge.*

▼ *No sign of damage to this sump, but I dressed up the edges after cleaning to make sure I didn't get any leaks.*

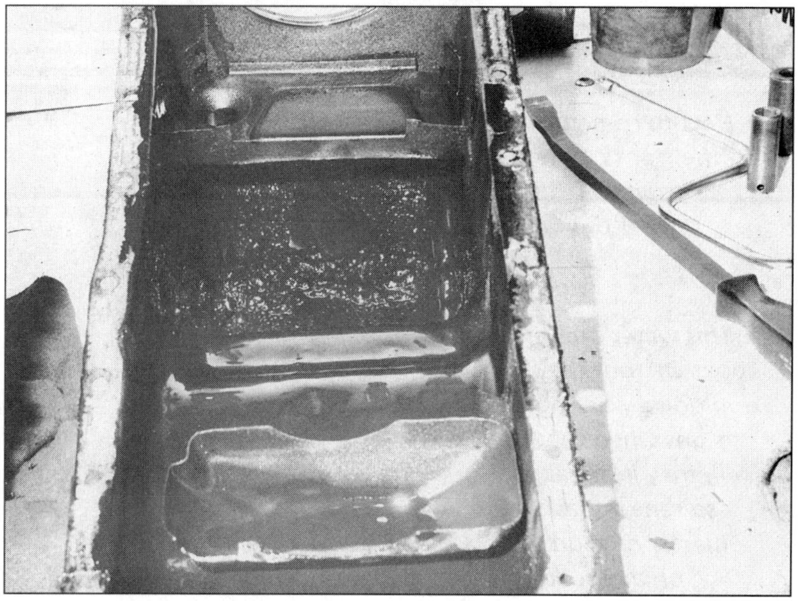

▶ *The inside of the engine looks reasonably clean – a sign that the oil has probably been changed fairly regularly.*

▶▶ *The first thing to come off from the bottom end is the oil pick-up pipe. It's held by a single bracing strap.*

▶ *The pick-up pipe just pulls out, but it might (like this one) be very tight because it was assembled with Loctite to make sure there weren't any air leaks.*

▶▶ *Round to the front now to take the timing cover off. Put a block of wood under the engine to save the crankshaft hitting the bench.*

◀ *Timing chains which have been running slack, particularly on Crossflow Fiestas, can score the inside of the timing cover. In severe cases they can cut right through.*

◀◀ *Remember that the dished oil slinger on the crankshaft fits with its concave side outwards.*

◀ *The chain tensioner is held by a single bolt.*

◀ *Swing the snail cam of the tensioner up and lift it off its peg on the front bearing cap.*

▶ *The face of the tensioner will be worn, so get a new assembly.*

▶▶ *The bolts holding the camshaft chainwheel are locked with locking tabs.*

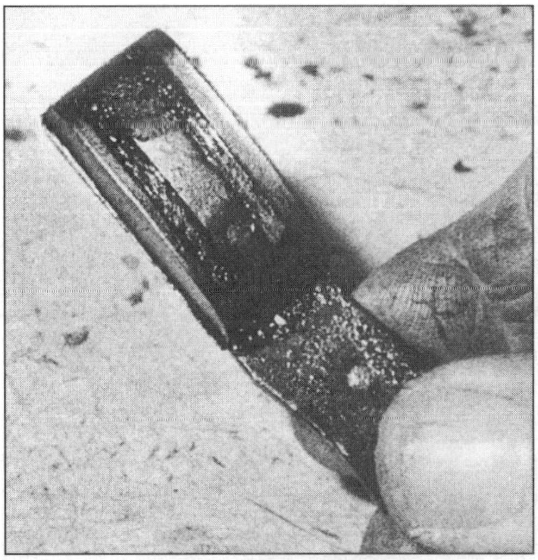

▶ *You may have to stop the crank shaft turning when you undo the chainwheel bolts.*

▶ *Use the pry bar to ease both chainwheels off, complete with the chain.*

▶▶ *More locking tabs hold the camshaft thrust plate bolts. Remember to get new ones.*

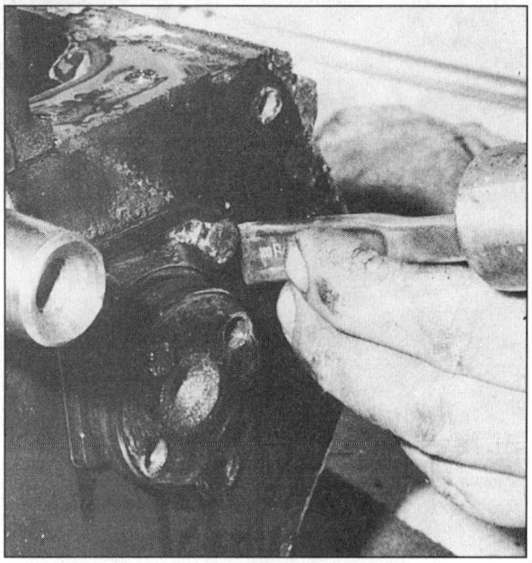

◀ *Undo the bolts and examine the thrust plate. If it's worn, get a new one.*

◀ *Give the camshaft a turn to make sure all the followers are free and lift it out of the front. Be careful not to catch the sharp edges of the cams on the bearings.*

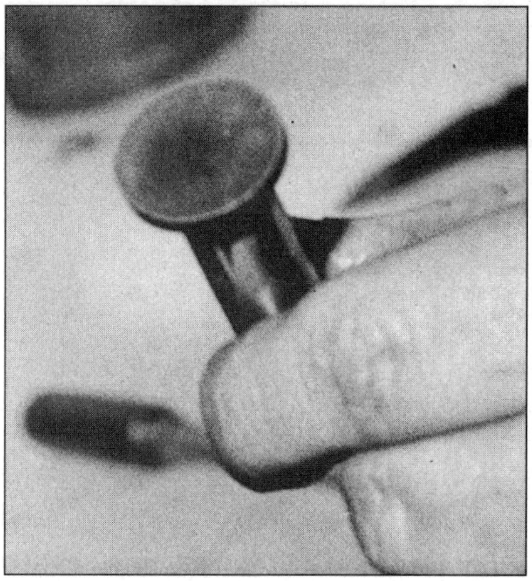

◀◀ *A rod with a magnet on the end is a great help in lifting out the cam followers.*

◀ *The faces of the cam followers should be slightly convex but they always wear flat. You must use new followers if you use a new camshaft.*

▶ *Two types of followers were used on the Crossflow – early engines had the ones with the slimmer stems, and the wider stems came in on 711M engines.*

▶ *After taking out the big-end cap bolts give the caps a light tap to free them.*

▶ *The rear oil seal housing usually needs a pry bar to get it off.*

▶▶ *The main bearing cap bolts will be tight and you'll need a long socket bar to undo them, maybe even with an extension.*

◀ *Have a look at the bearing shells as they come out. They can give you an indication of the state of the crankshaft journal bearings.*

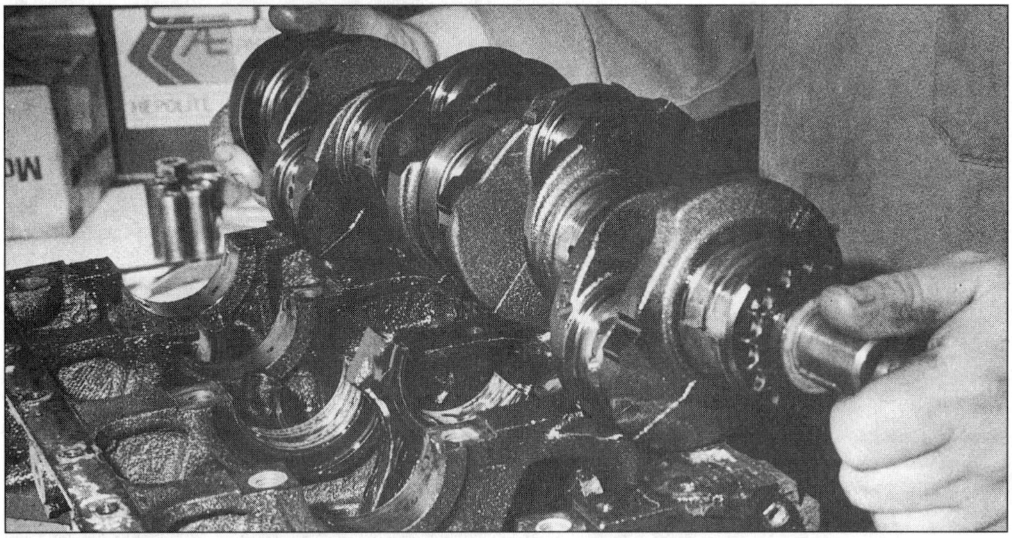

◀ *Lift the crankshaft out carefully, it's quite heavy and you don't want to drop it and bend it – yes, they do sometimes bend if they're dropped.*

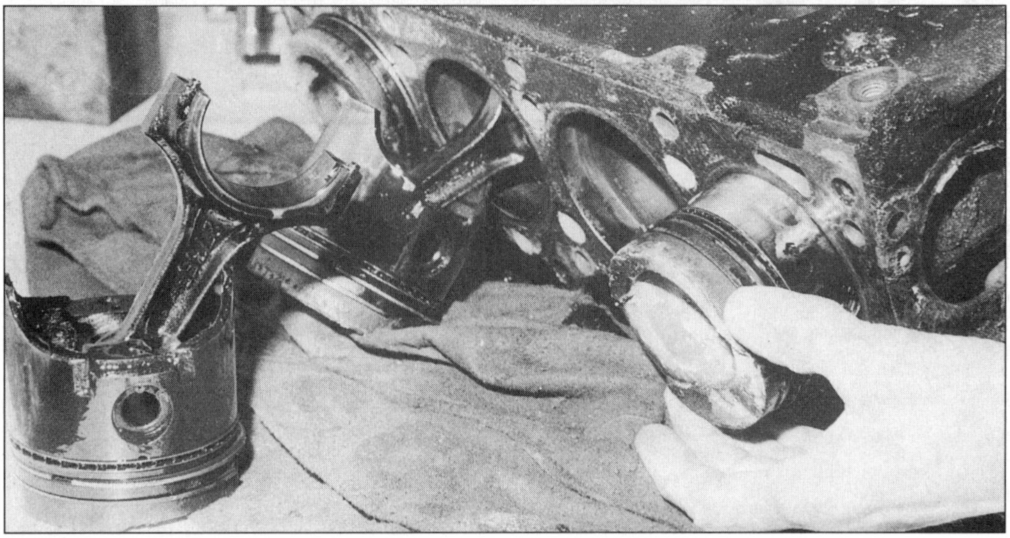

◀ *Tilt the engine on its side, push the con rods down and pull the pistons out from the top of the block.*

▶ The easy way to take out Ford core plugs is to drive one edge in with a flat drift.

▶▶ Driving one edge in tilts the core plugs over.

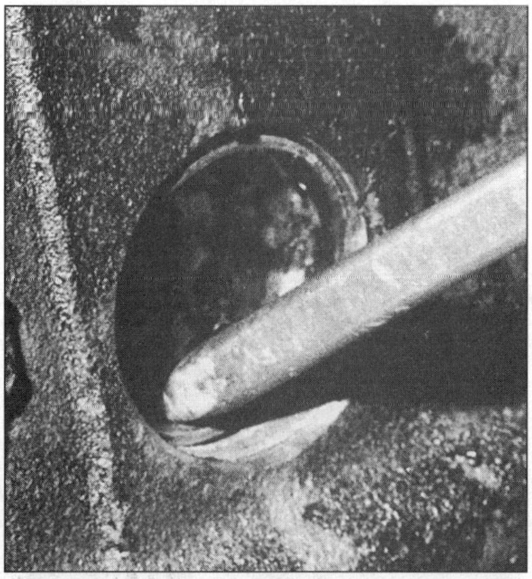

▶ Once the core plugs are tilted you can lever them out quite easily.

▶▶ Even core plugs which look good from the outside can be starting to rust through from the inside.

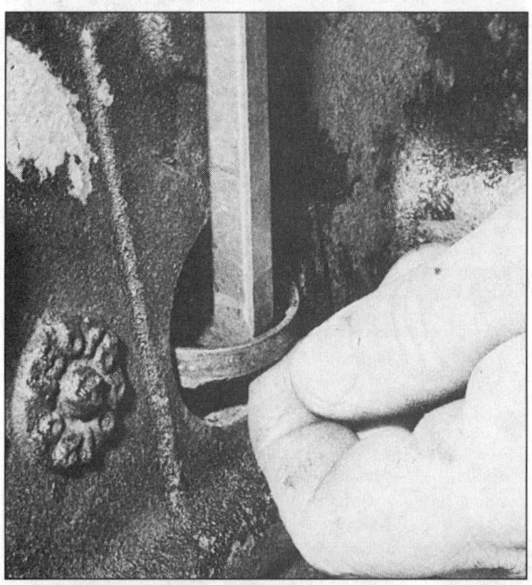

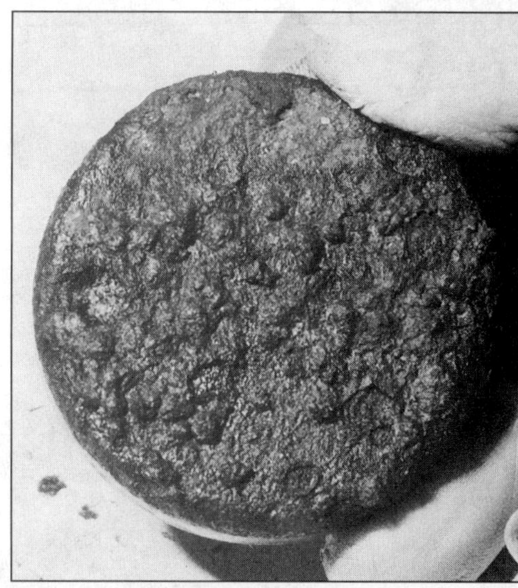

▶ Dig down through the water jacket with a long screwdriver to scrape as much sediment free as possible.

▶▶ You'll be surprised how much sediment there is in the water jacket once you start scraping.

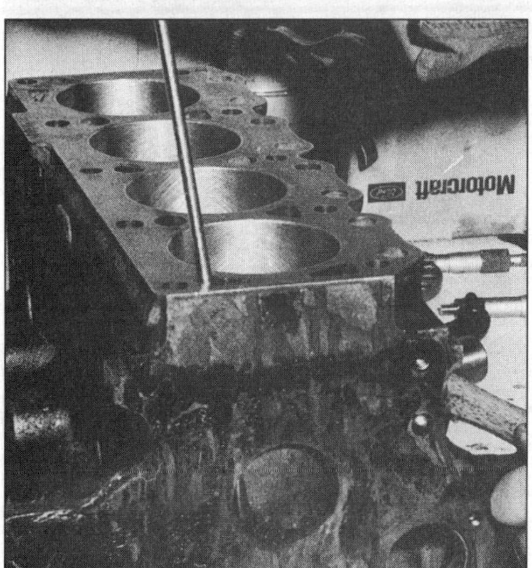

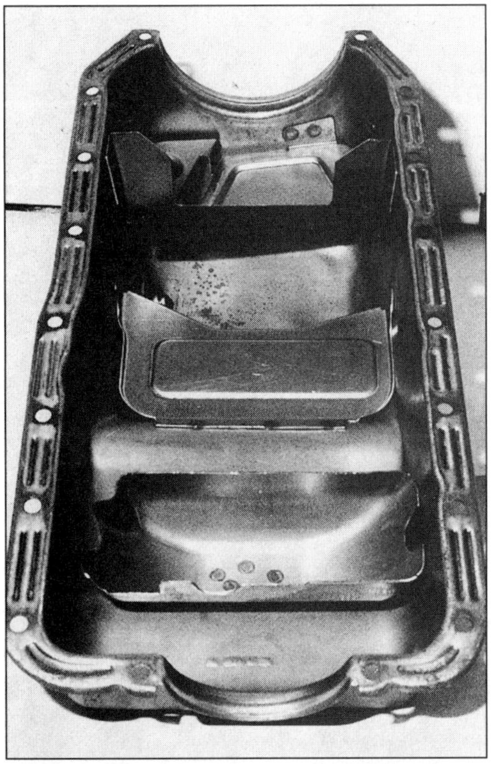

◀ *Check the joint faces of the sump and half-round oil seal grooves after cleaning, and check that none of the baffles has come loose.*

▼ *After washing the pick-up pipe thoroughly, give the strainer end a good blast of compressed air.*

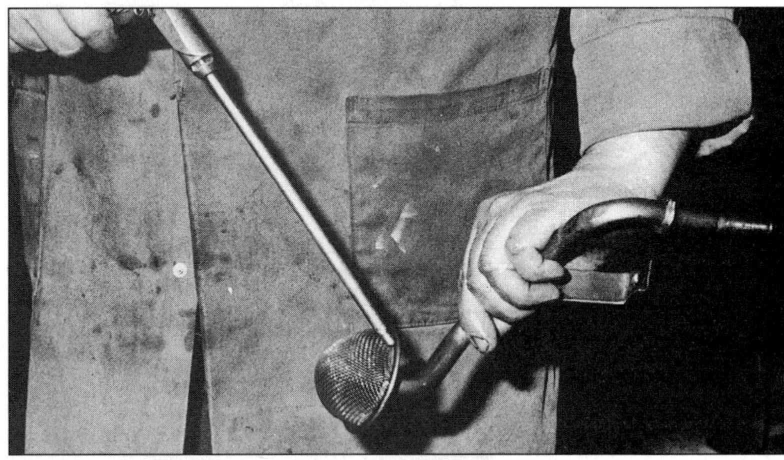

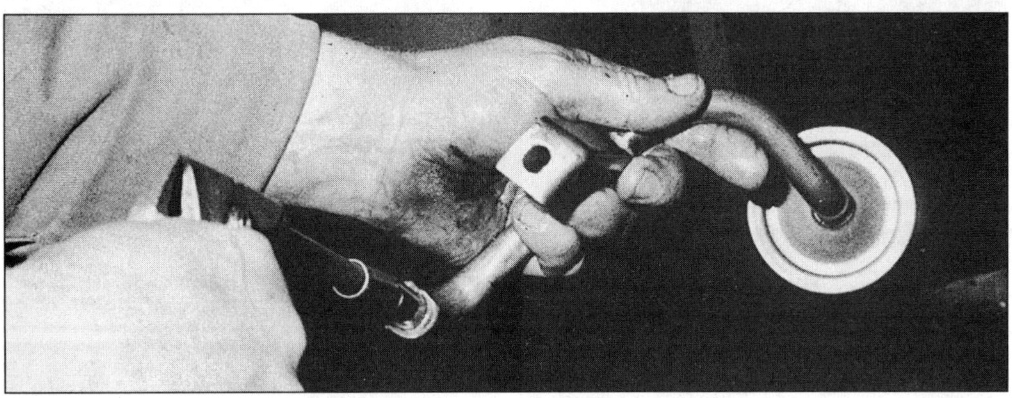

◀ *Give the other end of the pick-up pipe a good blow as well.*

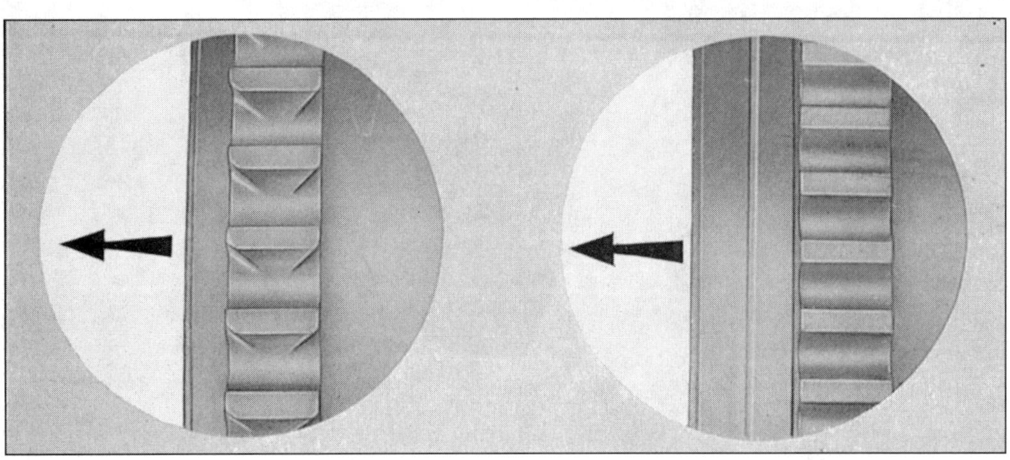

◀ *If you have to have the starter ring gear on the flywheel renewed, remember that there are two types. The one on the left (with chamfers to the teeth) is used with inertia starters. The one on the right (with no chamfers) is used with pre-engaged starters.*

▶ *Early Crossflows used a simple strainer pipe for crankcase ventilation.*

▶▶ *On most Crossflows you'll find this type of breather which can be taken apart for cleaning.*

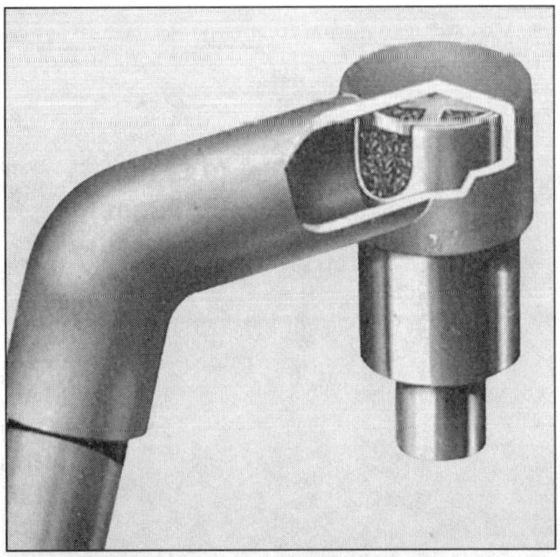

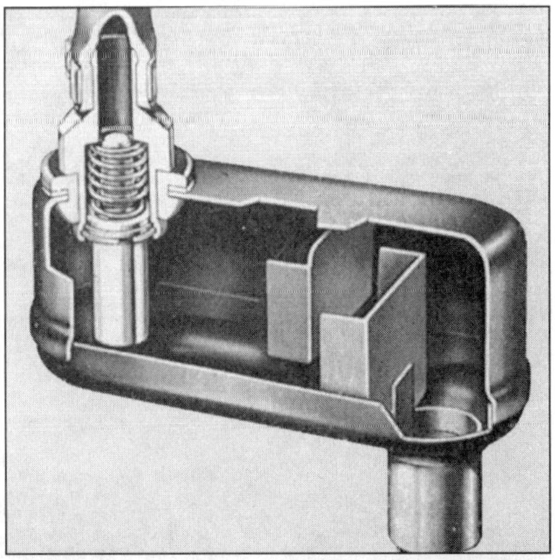

▶ *Before the block goes in the cleaning bath, take the Allen-headed plugs out of all the oil galleries so you can squirt through them with a syringe.*

▶▶ *If possible, send at least the block, head and crankshaft to be cleaned in a hot dip tank.*

▶ *The hot tank will get the parts much cleaner than you can at home, with much less effort and mess.*

◀ *When the block comes back from the cleaning tank, give all the water passages a blast of compressed air to drive out the sediment which the chemicals have loosened.*

▼ *Blow compressed air through all the oil passages, or use paraffin and a syringe, to make sure they're clean.*

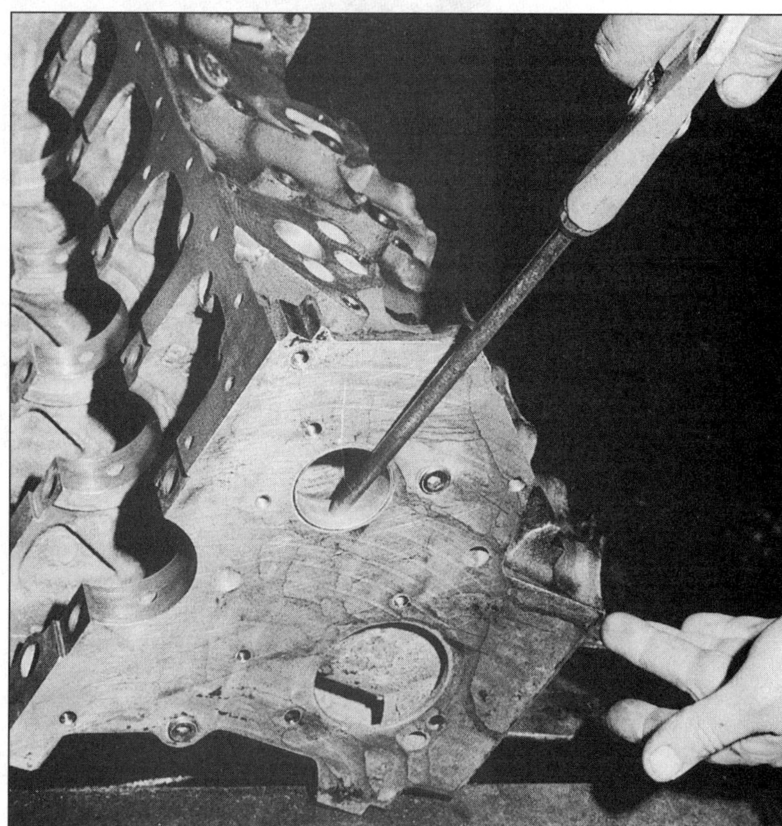

Chapter 8

The cylinder head

Over the years, the cylinder head and the gas flow through it has received more attention from tuners than any other part of the engine because it has the greatest single effect on the power output. As it stands, the gas flow through the Crossflow's head isn't as bad as some, but it can be much improved by tuners who know what they're doing and who have proved their theories by constant checking on a flow bench.

Conversely, it can be worsened by well-meaning but ill-informed and inexperienced mechanics. My advice on home gas-flowing is to limit it to matching inlet and outlet ports to the manifolds – going no deeper into the ports than about three quarters of an inch.

The vast majority of Crossflow heads you come across will be flat on their underside. The combustion chamber is all in the piston crown – a design known as the Heron combustion chamber. Some of the early engines, used mainly in Cortinas and Transits, also had a small combustion chamber in the head itself, but these heads are not generally reckoned to be the best for tuning.

Quite different are the special chambered heads produced by tuning firms like Vulcan Engineering. The Heron design with the combustion chamber in the piston has a number of advantages, but it also has disadvantages because it needs a heavy piston with much of the metal concentrated above the gudgeon pin. This will be dealt with more deeply in Chapter 13. For the moment, we'll stay with the more usual flat-face head.

Before you can consider home work on the head or putting it out to a tuner, or even if you are just overhauling the head with either no tuning or very mild tuning in mind, you have to make sure that it's in sound condition. Unlike the block, which is a massive casting by comparison and seldom warps much on its face, the Crossflow head does tend to bow unless the head bolts have been torqued down properly. Unfortunately, because the Crossflow is such a popular DIY engine, the chances of the head having been taken off and replaced by home mechanics, few of whom are likely to own a torque wrench, are high.

Before you start cleaning the face off, look carefully at both the gasket and the face of the head. You can often see signs of bowing by the spread of corrosion or rusty powder markings round the water passages, and blow-by marks round the top of the cylinders. Look round the firewall on the gasket where it fits round the cylinders. Any signs of burning are a sure indication that the head is warped. Check it again with a straight edge after you have stripped and cleaned it.

If you are just overhauling the engine

without reboring, you can get away with very slight distortion, and I mean very slight – like 0.005 in (0.130mm) or less – but, if you are reconditioning the engine back to standard with rebored cylinders, new pistons, reground crankshaft and so on, take it to a machine shop and get it refaced. If you have tuning in mind, get it refaced anyway, even if it appears to be perfectly flat. The extra pressures in a tuned engine will seek out even the smallest weaknesses.

Crossflow heads don't very often crack, but it's possible to crack them if the engine is allowed to overheat badly, and an almost certain way to crack them is to pour cold water in an overheated engine after it has lost its coolant. This may have been done by a previous owner who thought he had got away with it. I've seen one or two heads which had cracked across one of the exhaust valve seats even though the owners said the engine was running well.

Sometimes, a crack in a head is difficult to see until you have cleaned it thoroughly and gone round the valve seats with emery paper. If you have any doubt, take the head to be crack-tested. The usual type of machine which is used to check for cracks magnetizes the head so that any cracks become north and south poles of a magnet. The head is than painted or dusted with a solution of iron filings which cling to the gap between the north and south poles and highlight it. If you have tuning for high power in mind, I would advise having any head crack-tested as well as refaced.

Some people will tell you that a sure sign of a cracked head, or a blown head gasket because of warping, is a creamy deposit, like a mixture of oil and salad cream, inside the rocker cover, and that if you don't find this the head's all right. Like most saloon-bar pundit generalizations, this is a half truth and a dangerous one to rely on.

It's true that a creamy sludge is usually caused by a mixture of oil and water emulsifying, and that the water probably came from the coolant, but it isn't an infallible guide. Certainly it should be treated with the gravest suspicion, because Crossflows in which oil changes have been badly neglected and the breather has become clogged usually develop a thick black sludge. But I have known Crossflows that developed a creamy sludge with a perfectly good head, mainly because they were used for years on short town journeys and never reached their proper operating temperature for very long. The water vapour, which is naturally present inside the engine, condensed inside the rocker cover as the engine cooled down quickly in winter and formed an emulsion with the oil.

Conversely, the pouring of cold water into a hot engine treatment mentioned earlier could cause even the best maintained engine to crack at its hottest parts, the area round the exhaust valve seat. If the crack extended to the water jacket but not to one of the oil drain holes, you wouldn't get a creamy sludge, you would get an acrid smell of exhaust gases in the coolant and probably coolant loss and overheating as well.

Some of the 'sure tips' you may be offered by well-meaning friends who know that you are looking for a good second-hand engine to rebuild and tune may be perfectly sound but they should always be treated as positive signs that something may be wrong. Never accept the absence of symptoms as a sign that all is well. The only way to be certain is to strip the engine and check.

Some books on engine tuning go into great detail about gas-flowing the ports and the combustion chamber. As already mentioned, with two exceptions, which I cover in Chapter 13, the combustion chamber on the Crossflow isn't in the head, it's in the piston bowl so, from a gas-flowing point of view there are only the ports to worry about.

The theory of gas-flowing, and the experience of tuners who modify and test again and again till they achieve the best figures, is fascinating stuff to read but it isn't a great deal of practical use to the home tuner because it isn't an exact science and it takes a lot of experience and testing on identical heads to be able to wield a grinding stone inside a port with a guarantee of improving the gas flow. You also need to be able to get all four inlet ports and all four exhaust ports matched as closely as possible to have a smooth-running high-powered engine.

Repeating what I have already said in Chapter 3 – the old theory that you should get the walls of the ports as smooth as possible (even polished to a mirror-finish) has now gone right out of the window as it has been shown conclusively that a shiny-walled port doesn't give the best results since the fuel droplets tend to drop out of the mixture and collect on the smooth walls until they are large enough to fall back into the air-stream. This both upsets the flow and gives an uneven mixture. To prevent this happening, the walls of the ports should be left with a slightly rough finish from the grinding stone. What happens then is that the fuel droplets which try to drop out hit one of the ridges in the surface, bounce back and are picked up immediately in the air stream, before they have a chance to collect and grow.

When he opens out the ports to increase the amount of gas flowing through, the tuner will keep in mind the size of carburettor or carburettors you are fitting. There's absolutely no advantage in opening out the inlet ports to suit the flow through a pair of Weber double-choke DCOEs if you are going to fit a single double-choke DGV. Indeed, it would do more harm than good because the larger bore of the port would slow down the reduced flow of gas.

As well as opening out the port and improving the cubic feet per minute of fuel and air mixture that will pass through it, the tuner also tries to improve the swirl. The inlet ports on a Crossflow aren't in line with the cylinders. They are at an angle and this, combined with the shape of the throat just behind the valve seat, sets the mixture swirling round the bowl of the combustion chamber to get a more even burn. Swirl isn't something you can measure as easily as gas flow. Much depends on the skill and experience of the person who is grinding the port.

Keeping these factors in mind, we wouldn't advise anyone to try to improve the ports at home beyond matching them to the inlet and exhaust manifolds as detailed in Chapter 3.

It's almost certain that the valve guides will be worn, but this isn't serious because they can be replaced quite easily. Unless they have already been replaced you won't be able to fit new valve guides at home because (as I have already mentioned) the original guides are cast in with the head. They have to be machined out and new guides pressed in.

Even if the head on your engine already has removable guides I would advise having them pressed out and new guides pressed in by an engineering firm, because the new guides will never be accurately concentric with the old valve seats, so the seats will have to be recut.

At one time, Ford dealers used to keep valve guide reamers to ream worn guides out 0.015 in oversize, and they used to stock valves with oversize stems.

The reamers are still listed, as are some of the oversize stem valves, in Ford's parts list but, in most cases, the computer comes up with 'stock unavailable'. It was a good idea when labour was cheaper than parts but the valve seats have to be recut afterwards and, with the high price of labour nowadays, it would probably cost as much as an exchange reconditioned head.

One of the reasons why the valve guides wear relatively quickly in a Crossflow is that they are short and offer less bearing surface, compared with some other engines, to resist the side thrust of the rockers. The answer is to make the replacement guides longer. You can't make them longer at the top because, with a high-lift camshaft, the underside of the valve spring cap would foul the top of the guide. The only way to lengthen them is to let them protrude down into the port.

Theory says that anything which protrudes down into the inlet or exhaust port will upset the gas flow. However, the theory of gas flow through a port which isn't straight and, what's more, varies in its diameter and section shape, is far from an exact science. Testing on a flow bench by more than one tuner has shown that, provided the boss of the old valve guide is ground out to give a smooth roof to the port, having a valve guide sticking down has very little effect on gas flow in terms of cubic feet per minute compared with a guide which is ground off flush with the roof of the port. The effect is even less if the part of the guide which protrudes is tapered.

What is much more difficult to check is whether the protruding guide sets up any flow eddies which affect the swirl of the gases in the combustion chamber. Some theorists argue that, if there are any eddies set up, there is a good chance they will help and not hinder the swirl. I haven't built pairs of heads for A and B testing and I'm certainly not going to enter into a long debate about the mathematical theory of gas flow.

All I can offer is the experience of Paul Blanchard, a much respected tuner of engines who runs Vulcan Engineering in west London and who has built some very hot Crossflows. Paul has built Crossflows using heads with as near identical gas-flow figures as it's possible to get with hand fettling of the ports, both with and without protruding guides. On a dynamometer there was nothing at all to choose between the power outputs whichever head was fitted. Paul argues that the longer wearing qualities of the longer guides far outweigh any possible drop in power output, and all the Crossflows he tunes now have them. I am happy to go along with this proof of the pudding.

While you're having the old guides machined out, have the area round the guide spot-faced to give you some room to adjust the valve spring height. A good tuning firm will know exactly how much to machine off once they know what valve springs you are using, and the lift of the camshaft and rocker ratio. You need this extra clearance with a high-lift camshaft to be able to shim the valve springs to give you optimum poundage evenly along all the springs without the springs being compressed so far that they become coil bound. Should this happen, and the rocker is still trying to compress the spring, something has to bend or break.

When you buy your performance camshaft you will either buy it as a kit which includes new valve springs and spring caps, or you will be offered springs and caps matched to the lift of the shaft. With some camshafts, most performance shafts in fact, the camshaft maker will advise double valve springs.

As I've said before, this isn't because a single spring can't be made strong enough – the second spring is quite light in comparison with the main spring.

The reason is that the coil springs have a natural bounce frequency. If they are compressed and released at a speed which matches this natural frequency there will be a sudden reversal of energy and the valve will bounce open again. This can have a quite disastrous effect on power output. The job of the second, lighter spring is to damp the natural frequency of the main spring so that it puts it outside any speed at which the spring is likely to operate. You will need special valve spring caps, machined to provide a seat for the second spring.

While you are having the valve guides replaced, the seats cut and new valves, you may as well go the whole hog and have the head converted for unleaded fuel. As mentioned in Chapter 3, replacing the lead compounds in fuel by oxygenates has brought the octane rating of super unleaded fuel up to or even slightly higher than four-star leaded fuel. The only reason for retaining lead in petrol is to act as a lubricant to prolong the life of the exhaust valve and its seat.

If you choose a better quality steel for the exhaust valve than that used in the standard Crossflow valve, the only extra cost on top of the work already being done on your head is to have hardened exhaust valve inserts. If the original seats for the inlet valve were badly worn or have been badly ground by previous attempts at reconditioning, the tuning firm may advise you to have inlet valve seat inserts as well, but this is not a necessary part of the conversion to unleaded fuel.

Once you start tuning beyond something fairly mild, the big restriction to gas flow is the size of the valves. Ford's standard valve sizes are:

Engine capacity	Inlet valve	Exhaust valve
1100cc	1.42 in	1.24 in
1300cc	1.50 in	1.24 in
1600cc	1.56 in	1.34 in

These are nominal sizes, the actual measured size can vary because of production tolerances. All engine tuning firms offer conversion to valves with larger-size heads, with enlarged throats and ports in the cylinder head to match. You find variations on what is offered but, generally speaking, on Stage 1 heads the standard size valves are usually retained. Popular larger sizes on the 1600cc for Stage 2, sometimes called fast road tuning, are: inlet 1.60 in, exhaust standard. With higher stages of tune for competition, the inlet can go up as large as 1.72 in and the exhaust 1.42 in. Similar increases are offered on the 1110cc and 1300cc engines. Again, these are nominal sizes.

It might seem strange that the exhaust valve is left at its standard size on fast road engines and increased by only 0.80 in on a highly tuned competition engine, whereas the inlet valve is increased by twice that amount. Allowing for the already larger inlet valve, the difference in increased port area between inlet and exhaust is considerable.

The reason isn't hard to understand and it is why inlet valves are always larger than exhaust valves. With open ports, the only force pulling mixture into the cylinder is the depression below atmospheric pressure caused by the downward stroke of the piston.

On the exhaust stroke, we already have a mass of hot burnt gas at much higher than atmospheric pressure plus the pumping effect of the piston to drive the burnt gases out. A reasonable yardstick for a fast road head is that at the same pressures, a gas flow through the exhaust port around 80 per cent of that through the inlet port will give good scavenging.

Another dodge used by tuners to increase the flow through the inlet side is to thin the valve head behind the seat – the so-called 'penny on a stick' inlet valve. On a fast road head, most tuners use an inlet valve with a shallower cone to the back of the head than the standard valve, but it isn't wise to reduce the thickness of the head to the point adopted on some racing engines unless you go for a quite exotic steel.

At high revs with strong valve springs, the closing valve achieves a very high velocity by the time it hits the seat and there is a lot of momentum wrapped up in it. Reducing the mass of the valve by thinning the head reduces the momentum slightly, but there is still a tremendous shock loading when the head hits the seat. If the valve head is thinned too much with too small a radius between the back of the head and the stem, the concentration of stress from this shock loading introduces metal fatigue and the head of the valve can snap off.

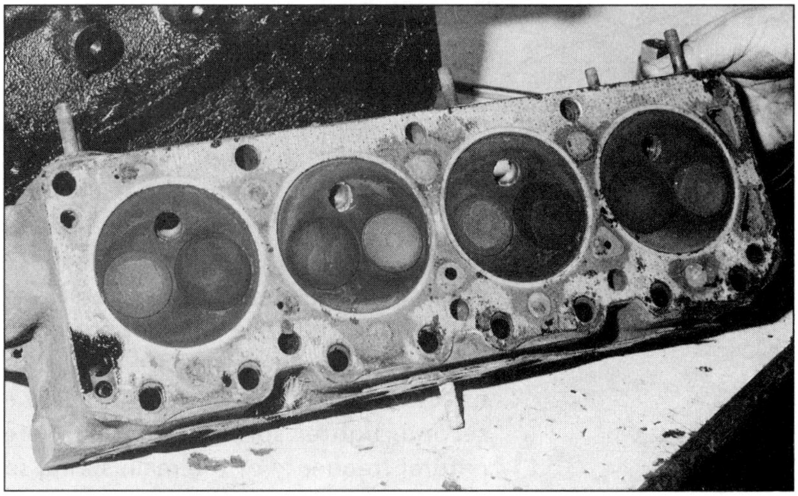

▲ *Most Crossflow heads will have a flat face like this one.*

On racing engines, the valves are usually changed on overhauls, which are frequent, so there is less time for fatigue to set in. Even so, it isn't unknown for inlet valve heads to snap off.

Some tuners like to put a small radius on the outside edge of the valve heads – inlet and exhaust – and many people think that this is also to help the gas flow. My view is that a small radius here makes very little if any difference to flow but could be a help in preventing pre-ignition by taking away the sharp thin edge that could glow red hot in the higher combustion temperatures reached in a tuned engine.

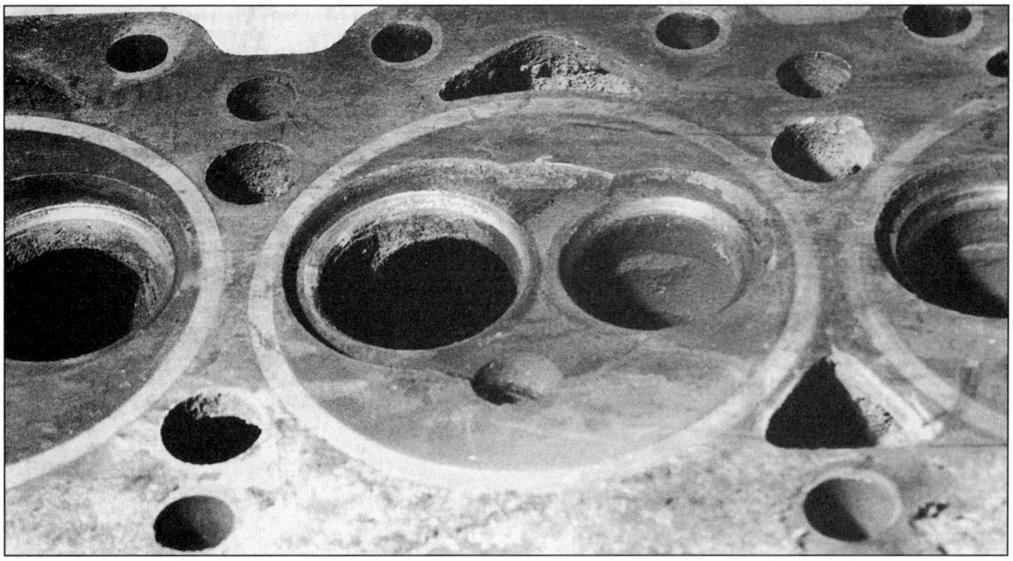

◀ *Some early Crossflow heads had a small combustion chamber in them. They are not generally recommended for tuning.*

▶ *Very different from the older heads are heads where a tuning firm has machined a combustion chamber so that they can be used with a flat-top piston.*

▶ *Almost all Crossflow heads will have to be refaced and the valve seats recut. This one has larger inlet valves.*

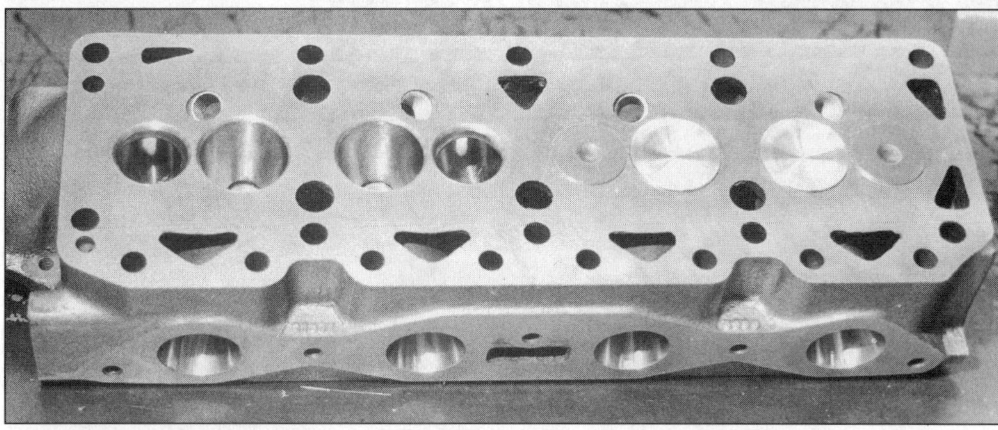

▶ *Full porting to improve the gas flow is a job for professionals.*

▶▶ *After gas-flowing, tuning firms check the heads on a flow bench.*

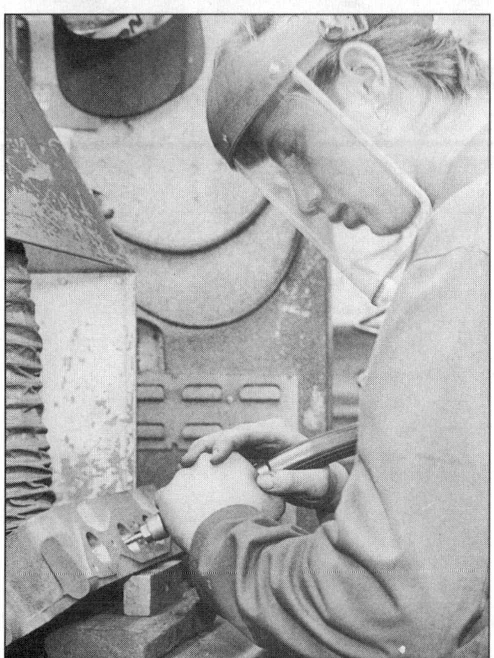

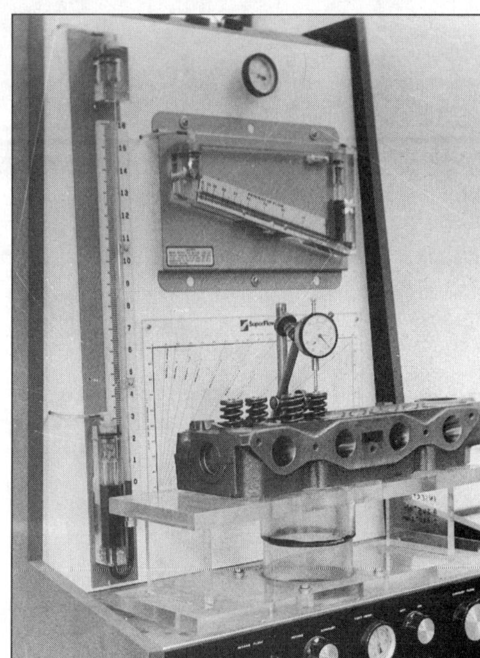

◀ *The most gas-flowing you should attempt at home is to match the ports to the manifolds.*

▼ *When the old valve guides are machined out, have the spring seats faced so you can use shims to equalize the spring poundage and use longer springs without the coils binding.*

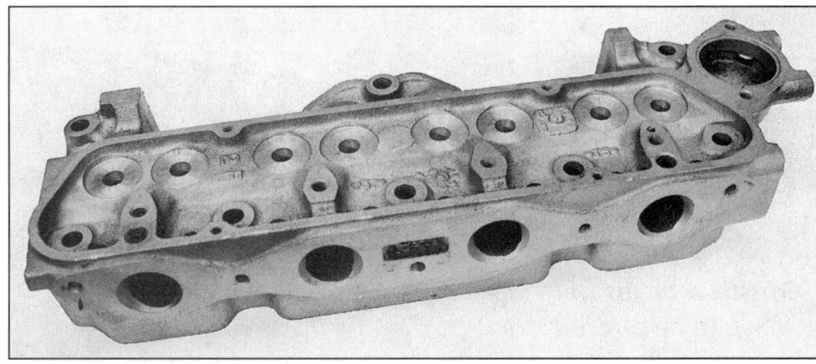

◀ *Most makers of performance camshafts advise double valve springs. Note the spring shimming to equalize the poundage.*

▼ *Ford valve seats have to be cut to three angles. A is the inlet seat and B the slightly smaller exhaust seat.*

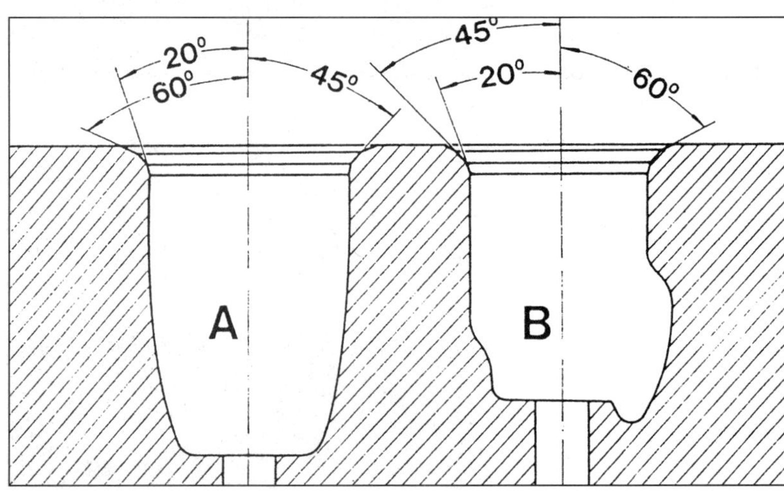

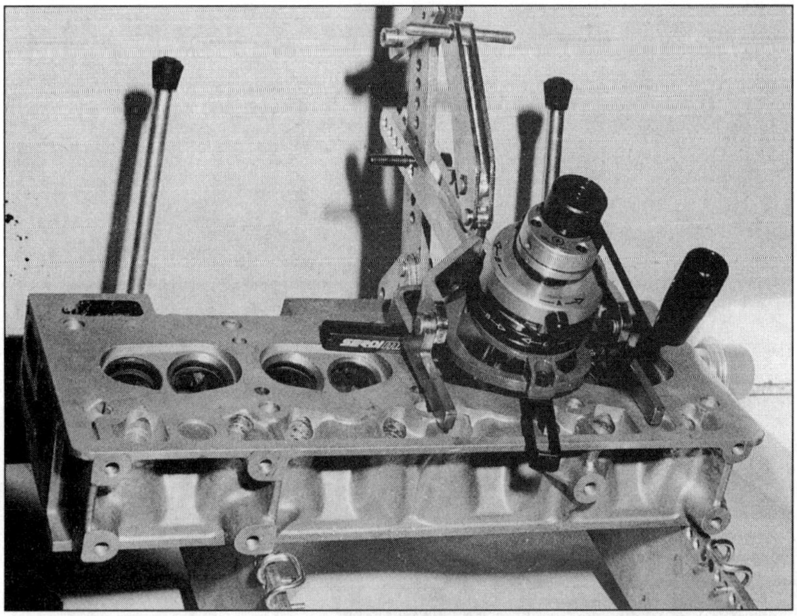

▲ *You can cut triple-angled seats at home with a portable cutter like this Serdi, but the equipment is expensive.*

▶ *Tuning firms will recut the valve seats with large accurate machines like this one at Burton Power Products.*

▶ *While you are having the seats recut, consider having the head converted for unleaded fuel, with hardened inserts fitted.*

▲ *After machining, tuning firms use jigs like this to adjust the spring poundage with shims.*

◀ *Many tuners recommend machining the back of the valve heads to give a better gas flow.*

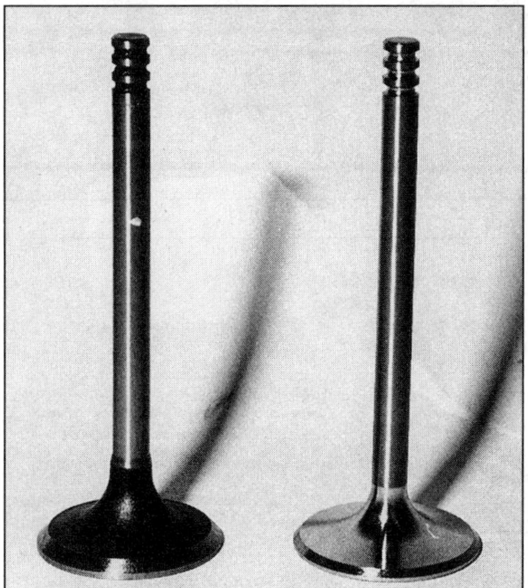

◀ *The difference between a thinned valve head, on the right, and a standard Ford valve.*

▼ *A gas-flowed head with new valves, seats and springs, ready to bolt on.*

Chapter 9

The valve gear

In this chapter on valve gear we are including everything from the camshaft to the valve stems: the cam followers, push rods, rockers and valve clearance adjusters, rocker shaft and pedestals. The valve gear is often a neglected part of engine rebuilding, particularly when home mechanics are just giving an engine a new lease of life, or rebuilding for a mild tune. Yet it plays such an important part in getting the best out of your engine, standard power or uprated, that it deserves more attention.

Starting with the cam followers, you'll probably find when you take them out that the base, the part which is in contact with the camshaft, looks flat, or

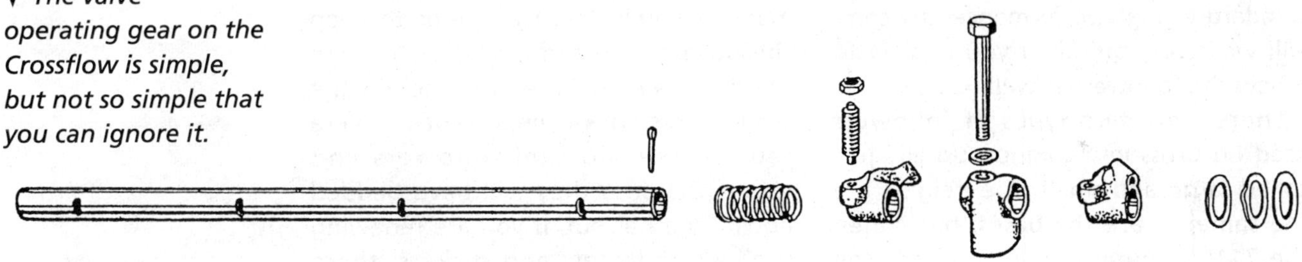

▼ The valve operating gear on the Crossflow is simple, but not so simple that you can ignore it.

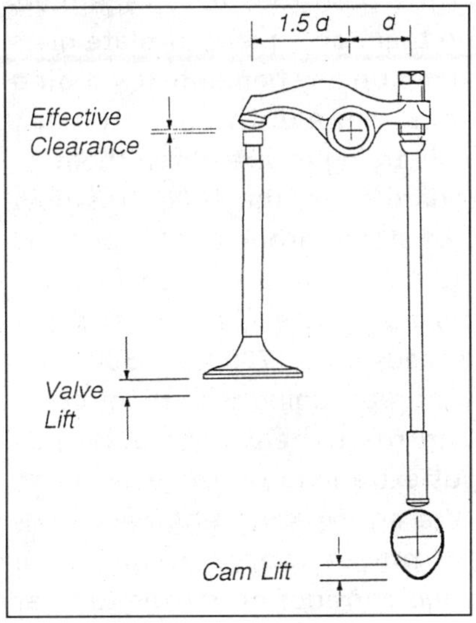

▶ The lift at the camshaft is multiplied by the rocker ratio, but the clearances have to be deducted to get the valve lift.

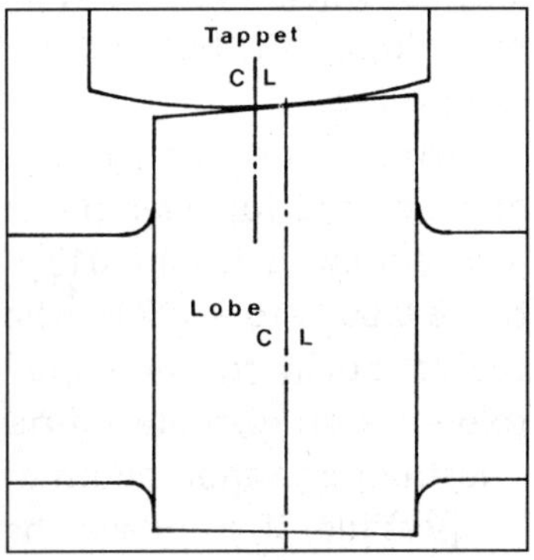

▶▶ The base of the cam follower is convex, the top of the cam lobe is sloped and the centre-lines are offset. These features rotate the follower to decrease wear and help to keep it free.

even slightly concave. It shouldn't be, it should be slightly convex.

The reason for this is to give rotation to the cam follower which lessens the wear on the base and also helps to prevent it becoming sticky in its bore in the crankcase. Just making the base slightly convex won't rotate it, the cam follower is rotated by the camshaft because the nose of the cam isn't ground absolutely parallel to the camshaft. It has a slight angle, or taper and its centre-line isn't quite in line with the centre-line of the tappet, so the wedging action when it pushes on the follower pushes the follower round.

In time, the convex base of the follower becomes flat, so it stops rotating, and this increases the rate of wear so that it soon becomes concave. It also increases the rate of wear on the cam. If you fit a new camshaft, either standard or high performance, the cams will wear very quickly if you neglect to renew the followers as well.

There were two types of followers used on Crossflow engines, usually just called large and small. The heights are the same, as are the bases, but, when the 711M engine was introduced, the stems were increased in diameter to give a larger wearing surface and lessen the rate of wear. It's usually the stems of the followers which wear more than the bores in the crankcase, so it's unlikely that when you fit new followers – either large or small – you'll run into problems of a sloppy fit, but check them for fit just in case.

Ford's tolerance for the clearance between the follower and its bore on a new engine is from 0.013 to 0.050mm, that's 0.0005 to 0.002 in – half a thou to two thou in common parlance. The tolerance on diameter of new followers is just under a thousandth of an inch, so it's possible, if you have the chance, to do a selective assembly of followers in the block to get the best fit. Tuners who are going for top power and who have a stock of followers do it but, on a standard rebuild or when tuning for normal fast road work, it's hardly necessary.

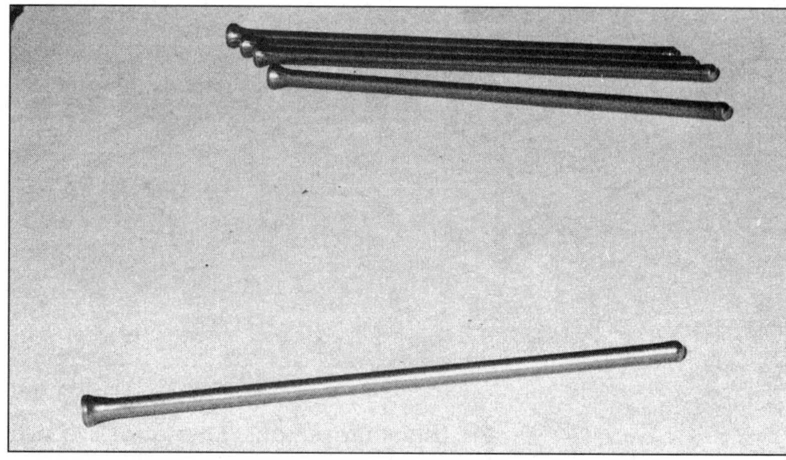

▲ *Check the push rods by rolling them on a flat surface, and reject any that are bent.*

Push rods very rarely give trouble and seldom show much sign of wear. In most manuals you'll see instructions to keep the pushrods in order when you take them out so that they go back in the same places. This is important if you're reusing the old cam followers and rockers because they will have bedded down together but, if you are renewing the cam followers and rockers, there really isn't a lot of point. It's worthwhile checking the push rods for straightness by rolling them on a piece of plate glass which shows up any bending. If you find bent ones, renew them – it isn't worthwhile trying to straighten them.

In standard form, the 1600 Crossflow ran out of steam around 5000 or 5500 rpm, and the standard pushrods are perfectly adequate for this and for engine speeds up to 7000 or 7500 rpm. If you want race tuning with high rocker ratios and really heavy valve springs which put extra load on the valve gear, and you're anticipating engine speeds of 8000 rpm or more, it might be worthwhile considering stiffer tubular pushrods. Some race tuners do and

▶ *The cast rocker shaft pedestals on this fast road engine have been replaced by steel ones to prevent flexing. The standard spacer springs are still used, but some tuners discard them and use spacer tubes.*

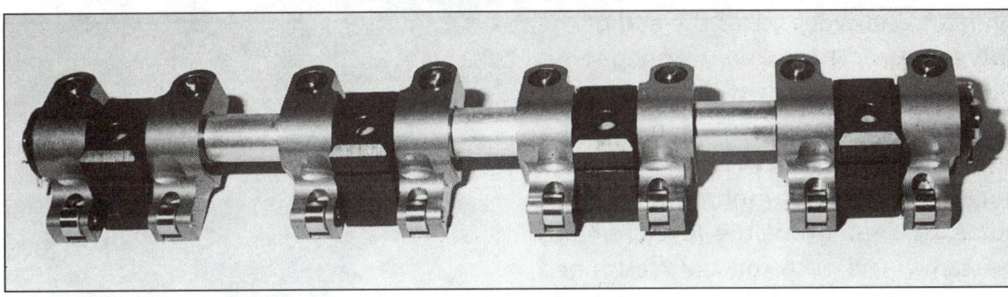

▶ *For racing power it pays to change to roller-bearing rockers, but they are expensive and of dubious benefit on a road going car.*

some don't. The latter stay with the standard rods as a sort of safety link, the philosophy being that if anything is going to bend it's better that it's a cheap and easily replaceable pushrod than anything more expensive and damaging to the engine.

Again, on a standard rebuild or a mild tune with only a free-flow air filter and performance exhaust, Ford's standard rocker assembly is quite adequate provided it isn't worn. It's almost inevitable that it will be, with a worn rocker shaft, worn rocker bushes and worn rocker pads. With any of these you won't be able to set the valve clearances to get a quiet engine and your power will definitely suffer, so renew the lot. If you're going for more power with a performance camshaft and gas-flowed head, the standard set-up isn't man enough.

Ford's standard rocker assembly locates the shaft in cast aluminium pedestals, with the rockers spaced along the shaft by coil springs. When you go for a performance camshaft with higher lift, possibly higher ratio rockers, stronger valve springs and higher engine speeds, two things can happen. The pedestals flex and, with weakened spacer springs, the rockers wander along the shaft against the springs, both of which upset the valve gear geometry and cost you power.

The answer isn't to fit a stiffer shaft, it's to use the standard shaft but with steel pedestals to hold it. New spacer springs are perfectly adequate for fast road tuning but, if you are going for competition work and feel like belt and braces, you can buy spacer tubes to fit on the unsupported parts of the shaft in place of springs.

We mentioned rocker ratio. If you look at one of the rocker arms you will see that the distance from the centre of the bush to the clearance adjusting screw is shorter than the distance from the centre of the bush to the centre of the rocker pad. In other words, the rocker is acting as a lever so that movement at the push rod end is multiplied by the ratio of the two arms of the rocker to give the valve increased lift.

The ratio of Ford's standard rockers is variously quoted as 1.52 to 1 or 1.54 to 1. This isn't because they vary, it depends on whether you take the maximum ratio or the mean ratio. As the rocker moves to open the valve the pad moves in an arc and slides across the end of the valve stem. This is why you get an elongated wear pattern on rocker pads instead of a round indentation. The end of the valve stem doesn't move sideways because it's restrained by the valve guide so one arm of the lever remains the same. The push rod isn't restrained so, as the adjustment end of the rocker moves in an arc, the effective arm of the lever lengthens and shortens.

At one time it used to be possible to buy rockers for the Crossflow with larger ratios to give greater valve lift. I haven't seen any ordinary rockers with different ratios about for some time now, but many of the roller rockers have a larger ratio – around 1.7 to 1.

If you are going racing and are looking for every ounce of power you can get, then large ratio roller rockers, expensive though they are, could be a good idea. With a road-going car, however, I would advise getting the valve lift you want from the camshaft and the standard rockers. The difference in friction is quite small compared with the other forces in the valve train, as is the difference in inertia between the standard cast rocker and an aluminium roller rocker. Some people fit them on road-going cars but, I strongly suspect, they do it because they want to copy the racing boys and not from any studied reasons of efficiency.

Set the valve clearances to the camshaft maker's specification for timing. They may or may not be the same as the running clearance.

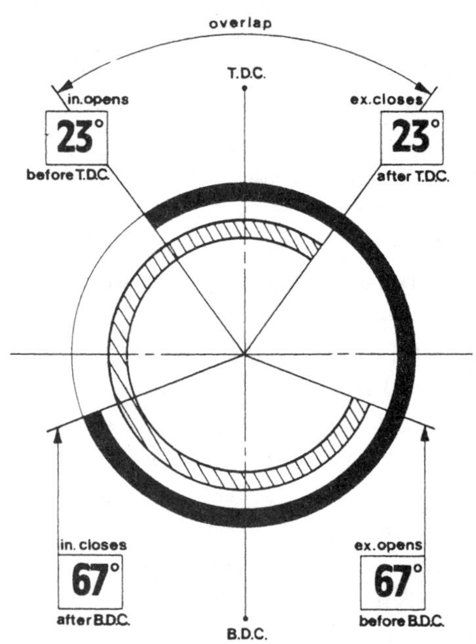

A typical valve timing diagram for a fast road cam.

Chapter 10

Camshafts

Because the camshaft on a Crossflow is down in the crankcase and not up on the head, fitting a new one means taking the engine out, so fitting a performance cam often comes after fitting a free-flow filter and performance exhaust. You should, however, consider the type of performance camshaft you are going to fit before you have the ports reprofiled and possibly larger valves fitted.

Camshafts for production family cars are designed for family motoring, where a smooth tickover, quiet valve gear, fuel economy and a low insurance grouping are the order of the day. Few family motorists worry about acceleration and response, and they couldn't care less what bhp their engine develops. This is why people like Kent Cams and Piper do a respectable if not roaring trade supplying performance camshafts to people who want more response and more power from their engines.

With camshafts it's definitely horses for courses – or horsepowers for course powers. If you are tuning for racing you'll want the ultimate the engine will take, and this means more than just a camshaft change, so you'll probably want to go to a specialist engine builder. You won't have to worry about emissions because there's no MoT for racing cars, but you'll worry about things like close ratio gearboxes to keep the engine inside a narrow rev band, and hang the fact that it doesn't tick over smoothly at low revs.

For road use you have to strike a compromise. You want the engine to be tractable so that traffic driving isn't more of a hassle than it has to be. You want decent acceleration in the mid-rev range to make overtaking safer and, let's be honest, for the sheer pleasure of putting a bit more zing and zest into your driving. You also want a reasonable increase of power at the top end so it doesn't take forever to overtake a truck on the motorway without changing down and, most important, you want to be able to sail through the MoT emission test.

These sound like very conflicting requirements, and some years ago they were, but camshaft design has advanced a lot since the days when you had to choose between a standard cooking camshaft and a wild stallion. With careful choice of the right camshaft you really can get the best of both worlds, even if insurance problems keep you down on cubic capacity.

First, let's lay the bogey that an engine with a performance camshaft is difficult to get through an emission test. Provided the rest of your engine – including the carburettor and the ignition system – are in good condition, there's no reason why, with a modern performance camshaft, you shouldn't get anything up to 30 extra bhp on a 1600 Crossflow with good response and with tickover emissions that will satisfy the most nit-picking MoT inspector.

Before we get down to what profile camshaft to choose, we'll have a look at some of the technical terms you'll come across. As well as lift and duration, you'll find camshaft men talking about base circles, opening ramps, opening flanks, noses, closing flanks, and closing ramps. You'll also hear talk of valve overlap and symmetrical and asymmetrical cams, so it's as well to get these all straight in your mind before you talk to camshaft suppliers and send for their catalogues.

We'll start with the base circle because the cam is built up on this. The base circle forms roughly 180° of the cam profile and doesn't impart any movement to the cam follower. You set the valve clearances when the follower is in contact with it.

Moving up the side of the cam we come to the opening ramp, which is closely related to the opening flank. The ramp's job is to take up the valve clearance and start the follower moving in a controlled way that doesn't put a sudden shock into the valve train. On a standard touring cam the opening ramp plays an important part in keeping the valve train quiet, but it does slow down the opening of the valve. When we are tuning for power, we want to get the valve open as quickly as possible, so on a performance camshaft the opening ramp is sharper.

Unfortunately this tends to make the operation of the valve gear more noisy. If you listen to an engine tuned to racing power, particularly if you use a probe stethoscope, the valve gear is usually very noisy. Any noise with moving parts means wear, and the greater the noise the greater the wear but, on racing engines, the tuner is

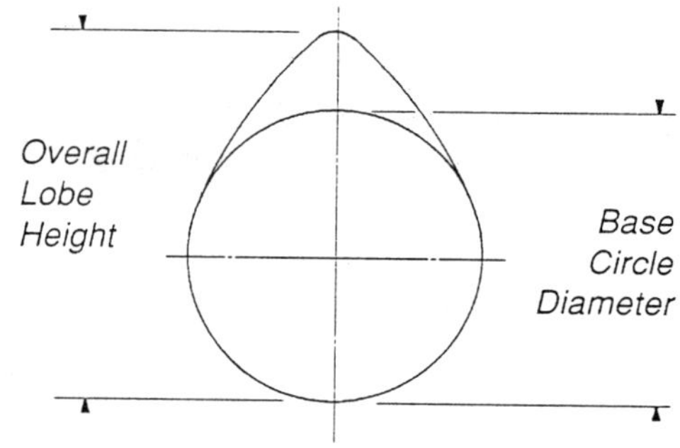

◀ *Cam lift is arrived at by deducting the base circle diameter from the overall lobe height.*

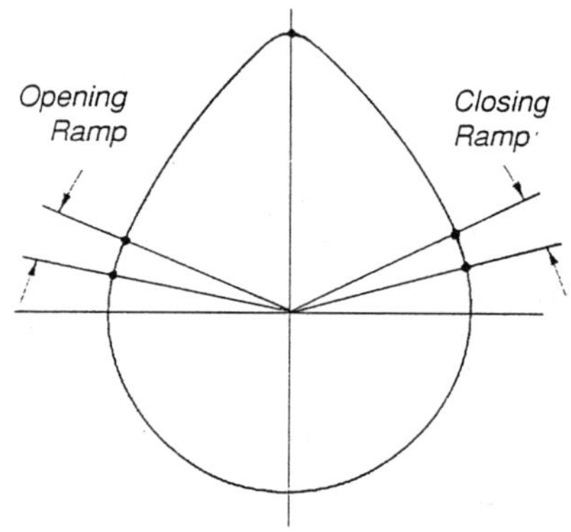

◀ *The opening and closing ramps lessen the shock of the cam follower starting to rise and coming to a stop.*

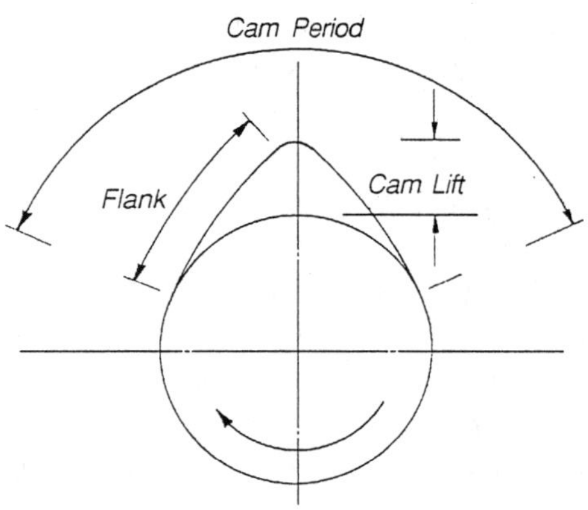

◀ *The cam period is the total time the valve is off its seat, from the start of the opening flank to the end of the closing flank.*

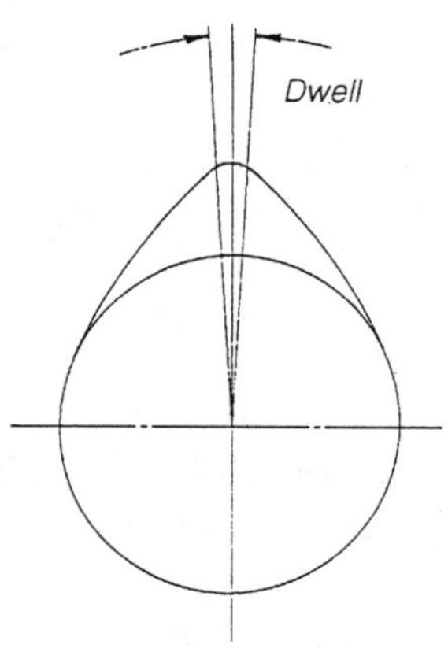

◀ *The valve has to remain still for a moment at full lift before it changes direction and starts to close. This is determined by the width of the cam nose, or dwell.*

▶ Talking about cam duration is meaningless unless you also quote at what clearance the duration is calculated or measured.

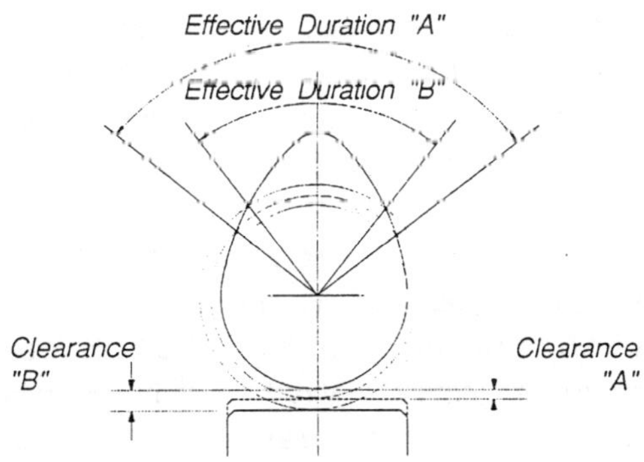

▶ Overlap is the period during which both the inlet and exhaust valves are open together. It is governed by the angle between the inlet and exhaust cams known as the lobe centre angle.

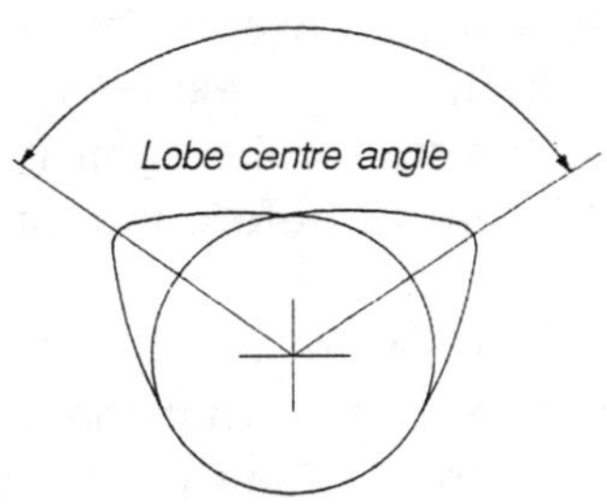

▶ This diagram shows how it is possible to increase the lift of a cam by grinding something off it. Reducing the base circle diameter increases the difference between it and the overall lobe height.

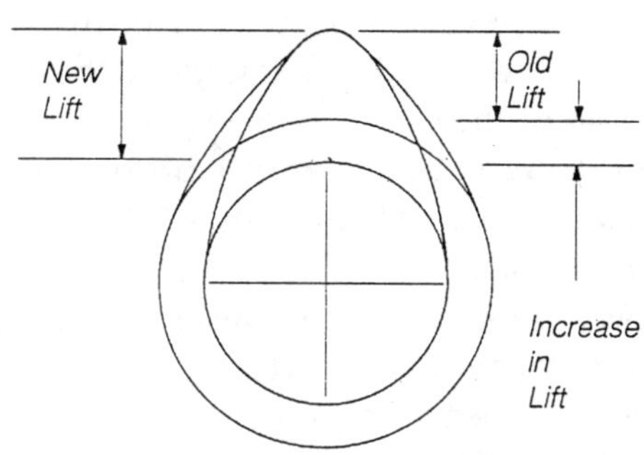

▼ You sometimes hear of asymmetric cams where the nose is offset to one side of the centre-line. They are usually used on engines with finger followers, like the Pinto, but very seldom on a Crossflow.

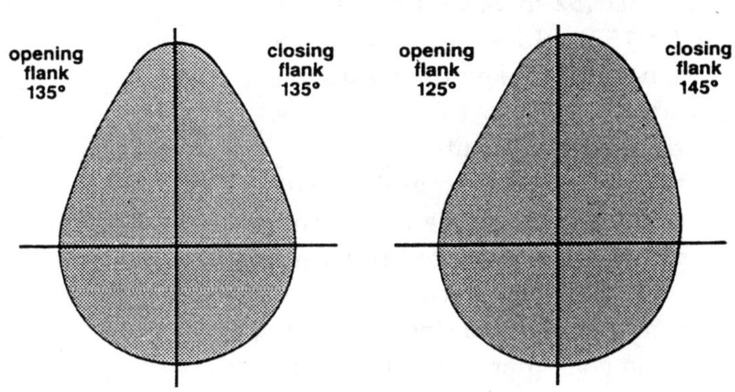

looking for every last ounce of power, and the engines are stripped and checked far more often than fast road engines – sometimes after only one or two races. On a normal fast road engine few people would put up with either the noise or the increased rate of wear, so the cam designer has to strike a compromise between fast valve opening, noise and wear.

After the opening ramp, cam follower movement is taken over by the opening flank which accelerates the cam follower and pushes the valve open to its full lift position. The opening flank has far more work to do than most people realize, and it is a critical factor in profile design. It has to accelerate the whole of the valve train – follower, push rod, rocker, spring cap, spring and valve – but it has only about half its length to do this because it also has to retard the follower before the valve reaches its full lift position. If it didn't, the follower would jump off the top of the lobe and bounce.

The nose of the cam, the pointed bit between the opening and closing flanks, is the point where we get maximum valve lift. It's also the part of the profile which is most highly stressed and which wears the fastest because it has the least area in contact with the follower.

Like people, different cams have different shaped noses. Some are quite pointed, some are rounded and some almost appear to be flattened at the tip. As well as giving the valve its maximum lift, the nose of the cam is also the part of the profile where the valve movement stops and changes direction. The length of time the valve remains fully open is called its dwell, and the angle through which the cam turns while the valve is fully open is the dwell angle. Just to confuse things, dwell angle isn't usually quoted in degrees of camshaft rotation, it's quoted in degrees of crankshaft rotation. As the camshaft runs at half engine speed on a four stroke engine, the dwell angle in crankshaft degrees is twice the dwell angle at the cam profile. Actually, this isn't just to confuse – the reason is that the dwell angle is important when you come to set up the camshaft timing and it's easier to use degrees of crankshaft rotation than camshaft rotation.

After the nose has held the valve open, we come down the other side on the closing flank which first accelerates then decelerates the follower till the closing ramp lowers it gently down on to the base circle again. The closing ramp is usually more gentle than the opening ramp because fast closing of the valve isn't anything like so critical for performance as fast opening.

The other point about having a more gentle closing ramp is to guard against the valve crashing back on its seat, which would do neither it nor the seat any good. Sudden crashing back could also cause valve bounce because of the sudden reversal of energy in the spring, and this would cause the valve to slightly open again just when it was least needed and would lead to misfires and burnt valve seating.

Next, lift and duration. Lift is fairly straightforward – it's the amount by which the cam lifts the cam follower. Here, you have to be clear whether you are talking about cam lift or valve lift. On an overhead camshaft engine with a direct-acting camshaft on top of the valves through bucket tappets, the valve lift is the same as the valve lift minus the valve clearance but, on a pushrod engine like the Crossflow you also have to take rocker arm ratio into account.

Cam lift, as quoted by the camshaft maker, is the overall lobe height minus the base circle diameter. Some people try to check this by using a micrometer or vernier gauge, but this isn't a very accurate way of doing it because you can never be certain where the base circle starts and ends. The chances of arriving at an accurate base circle diameter are very small.

If you must check what the maker says, or if you want to check the cam lift of an unfamiliar camshaft, the only way to do it accurately is to set it up in a lathe between centres and use a dial test indicator. You have to make sure that the DTI is set with its probe pointing exactly towards the centre of the camshaft, and the easiest way to do this is to set the probe against one of the lathe centres before mounting the shaft. Checking in a lathe also lets you check the run-out of the centre bearing journal to make sure that the camshaft hasn't been bent. Camshafts are pretty stiff, but they can bend if they're dropped on a hard surface.

Some people check the cam lift by using a vernier depth gauge at the top of the pushrod. It will give you a pretty good idea, but it won't be quite as accurate as checking with a DTI on the cam profile.

As a general rule, all pushrod engines have more valve lift than cam lift because the two arms of the rocker, the pushrod side and the valve side, are not equal, the valve side being the longest. As we've seen, the Crossflow is no exception, the ratio between the arms on a standard rocker being 1.52 to 1. So, for every 0.010 in cam lift we get a theoretical 0.0152 in valve lift. (0.010 x 1.52 = 0.0152). However, the valve clearance has to be taken up before the valve starts to lift, so this has to be deducted from the valve lift.

Let's take an actual example. Kent Cams' BCF2 fast road cam for the Crossflow has a cam lift, inlet and exhaust, of 0.261 in. Because of the design of its opening and closing ramps, Kent recommends that valve clearances of 0.014 in inlet and 0.018 in exhaust are used instead of the standard Ford valve clearances. So, with a rocker ratio of 1.52 to 1, the Inlet valve lift becomes (0.261 x 1.52) – 0.012, which comes to 0.384 in. The exhaust valve lift becomes (0.261 x 1.52) – 0.012, which comes to 0.378 in.

So far, so good – theoretically. Unfortunately, the practicalities of engine manufacture mean that components are manufactured within tolerances rather than to a specific figure using three decimal places. Lengths of cam followers, lengths of pushrods, ratios of rocker arms, heights of rocker pedestals, lengths of valves and the heights of the valve stems above the top face of the head can all vary slightly. Valve heights also alter each time the valve seat is recut.

Because the rocker pad which pushes the valve open works on an arc then, as a general rule and with the normal rocker angle of a Crossflow head that hasn't been wildly modified, if the valve height is reduced the effective rocker ratio will increase slightly to give more valve lift. Conversely, if the valve height is decreased the effective rocker ratio will decrease and we get less valve lift.

Sometimes, the standard production tolerances can all work in our favour to give us the best valve lift possible with any given set of followers, pushrods, rockers, pedestals and valves. At other times, all the tolerances can be against us and we get less valve lift. This is where a conscientious engine tuner can, by careful measurement and by juggling with various parts, produce an engine with a few more bhp power output than someone who just takes parts as they come and puts them together.

This is part of what's meant by the often over-worked term of a 'blueprinted' engine. This is an engine built to standard specification but with a power output above that quoted by the manufacturer. All the parts are to manufacturer's specifications and tolerances but, by a careful choice of tolerances that work in his favour, and by careful fitting and assembly, the tuner can increase the engine's performance. The same thing applies to an engine modified and tuned for performance. This is why manufacturers of performance camshafts, free-flow air filters, performance exhausts and gas-flowed heads always quote a range when they talk about the increased power you can expect, rather than give a specific figure.

You can work out the valve lift this way, but in this case it's easier and more accurate to measure it than calculate it. If you set a DTI up on the top of the cylinder head with its probe on the top of the valve spring retainer you can get a direct reading of the valve lift. It can often be quite interesting and instructive to measure the valve lift on all the valves and then compare it with the calculated figure.

Duration is nearly always quoted as cam duration but, in fact, it is measured in degrees of crankshaft rotation and is the number of crankshaft rotation degrees during which the valve remains off its seat. So if, for example, we have a cam duration quoted at 270°, this is 270° of crankshaft rotation. The camshaft runs at half engine speed, so in terms of camshaft rotation the figure would be 135 degrees.

Duration, like valve timing, ought to be quoted with a specified valve clearance. If, for example, you set the valve clearance to 0.020 in and measure the crankshaft rotation during which the valve is off its seat you will get a certain figure. If you now reduce the clearance to 0.010 in, the valve will open earlier and close later, so you get a new duration figure larger than the first.

You ought to keep this in mind when comparing camshafts from different makers, and ask what valve clearance was used to measure the duration. As a general rule, though not an infallible one, the longer the duration the less likely you are to get the engine to tick over smoothly at low speeds.

As a general rule, Crossflow engines don't like an over-long duration. Typical figures for performance camshafts could be 270 to 280° for a fast road camshaft, and up to 310 or even 320° for a full race shaft, but the latter would be almost impossible to drive in traffic because the engine would hardly run at all, let alone smoothly, below 2500 to 3000 rpm.

They do, however, thrive on lift, and that brings us to increasing the lift of a cam by regrinding the existing profile. Many people find it difficult to understand that you can increase the lift by grinding something off it. Initial thoughts suggest that this will decrease the lift.

If, however, you think of lift as overall lobe height minus the base circle diameter, it becomes clearer. If you alter the ratio between the two – in other words reduce the base circle diameter by more than you grind off the lobe – the cam follower sits lower on the base circle and the lobe will lift the cam follower by a greater amount than before.

You sometimes find on reprofiled cams that the parts between the cams, usually called the core, have also been machined. This isn't just for looks, it's because the base circle has been taken down lower than the core and, if the core were not machined as well, the cam followers would sit on this and not on the cam. A completely machined camshaft doesn't necessarily mean that it has been reprofiled. Some high performance cams are turned from new castings where the core has to be machined down to clear the cam base circles.

We said that Crossflow engines like quite a lot of lift, but you have to be careful. A few early Crossflows, mainly used in some Cortinas and Transits built before August 1970, have a very small combustion chamber of about 7cc in the head, just enough to accommodate the valve heads, as well as the chamber in the piston crown, but all the later ones have a completely flat head-face, with the combustion chambers in the piston crowns. Don't confuse these heads with the special chambered heads developed by tuning firms which are referred to in Chapter 8.

If you use too high a lift there's a danger that, when they are fully open, the valves will touch the tops of the pistons with disastrous results. Things can get even tighter if the head-face or the top of the block, or both, have been machined to take out bowing or warping, or if you have had oversize valves fitted. With many high-lift camshafts it's necessary to have the tops of the pistons machined to clear the valves.

With a block and head that haven't been machined, and with standard size valves, you ought to be able to rely on the camshaft maker to tell you that there is sufficient clearance. However, with a second-hand engine you can never be certain whether or not the faces have been machined so it's as well to check in all cases.

The easiest way to do this is to assemble the engine dry with a thin layer of Plasticine in the piston cut outs. Set the valve clearances carefully to their running clearance. Then rotate the engine slowly and stop immediately if you encounter any hard resistance. If everything seems clear, take the head off and measure the thickness of the Plasticine where it has been squashed down by the valve heads. If you find a clearance less than 0.060 in (1.52mm) – 0.080 in (2mm) is better – it's unsafe at high revs. You also ought to allow at least 0.040 in (1mm) sideways clearance between the valve heads and the sides of any cut-outs in the piston crowns. Theoretically, the valves won't touch the pistons if there is only a small clearance, but parts flex and valves can bounce at high revs no matter how strong the springs or how many you use. Play it safe and make sure you have adequate clearance.

Valve clearance cut-outs in pistons may be necessary, but they aren't the most desirable things to have from the point of view of effective combustion. Any pockets in a combustion chamber tend to harbour mixture and this can cause detonation. See Chapter 13 for what to do about it.

Other things you have to be careful about with high lift are valve spring coil clearance, and clearance between the valve spring retainer cap and the top of the valve guide. With your new performance camshaft you will get a set of new valve springs, invariably stronger than the standard Ford ones, which will have been matched to the camshaft. They won't, however, be matched to your engine because the camshaft maker has no way of knowing what you've done to the head.

This is why, when you fit a high lift cam, there is a distinct danger, if you are not having any other work done on the head, that the valve spring will become coil bound – that is, compressed to the point where the coils touch each other and the spring goes solid. Should this happen before the camshaft has finished lifting the valve something has to bend or break.

This is why, when you have the old valve guides machined out and new ones pressed in, you should ask for the valve spring seats on the head to be machined down. Then you can set your valve springs up with shims to give you maximum poundage without any danger of the coils binding. Set the shims so that you can insert a 0.040 in (1.0mm) feeler gauge comfortably between the middle coils when the valve is fully open.

Before that you should have checked the clearance between the underside of the valve spring retainer cap and the top of the valve guide. The easiest way to do this is to assemble the valve with a light spring in place of the proper valve spring. An old rocker spacing spring from a Crossflow's rocker shaft is ideal. With the valve fully open, there should be a clearance of at least 0.060 in (1.52mm) between the under-side of the retainer and the top of the valve guide. If there isn't, reduce the height of the guide.

To give you an idea of the lift differences we are talking about between a standard camshaft and a high lift one, Ford's standard camshaft on a 1600 Crossflow gives an inlet valve lift of 0.347 in. On a fast road cam it is usually in the region of 0.380 to 0.382 in, and on a full race cam 0.490 to 0.493 in.

If you look at a valve timing diagram you'll see that the inlet valve opens before top dead centre and the exhaust valve closes after top dead centre. The number of degrees of crankshaft rotation during which both valves are open at the same time is termed the valve overlap.

There are a number of reasons for arranging things this way. The inlet gases have inertia so they take a short time to get moving. Before top dead centre the piston is still on its way up, so you might think that if the inlet valve opened before top dead centre the inlet gases would be pushed back in the manifold. However, the exhaust gases rushing out of the exhaust port create a suction effect inside the combustion chamber so that the inlet gases actually start moving in even though the piston is still moving up.

The piston hasn't expelled the last remnants of the exhaust gases by the time it gets to top dead centre and, if there were no incoming inlet gases, then if the exhaust valve were still open, the remaining exhaust gases would get sucked back in as the piston moved down. However, the inlet gases are well on their way by this time, and before the exhaust gases have time to slow down and change direction the inlet gases have pushed most of them out of the exhaust port.

You can see that the amount of valve overlap is a critical factor in engine performance. If you have too little, a lot of exhaust gas would be left inside the combustion chamber so there would be less room for the incoming charge. Too much, with the inlet valve opening too early, and you would get exhaust gases pushed into the inlet manifold. Quite apart from the danger of the hot exhaust gases igniting the mixture in the inlet manifold and giving a backfire, the mixture – when it finally changed direction and came in – would contain exhaust gases as well as new mixture so the power would be down.

Too much overlap, with the exhaust valve staying open too long, would mean that some of the new mixture would escape out of the exhaust port and probably be ignited in the down pipe. In extreme cases you could get sheets of flame out of the tail pipe.

Valve overlap is governed by the positions of the inlet and exhaust cams relative to each other. If you look along a camshaft from the front you will see that there is an angle between the inlet and exhaust cams. This is called the lobe centre angle. The smaller the lobe centre angle, the greater the overlap.

Timing Chain and Sprocket

As standard, the Crossflow uses a single-row chain for driving the camshaft. For a standard production engine this is quite adequate and, indeed, I have never come across problems with a single row timing chain fitted to a fast road engine, even after 50,000 or 60,000 miles. There's certainly no danger of it breaking but, people will tell you, it can stretch.

We agree that, when a chain wears,

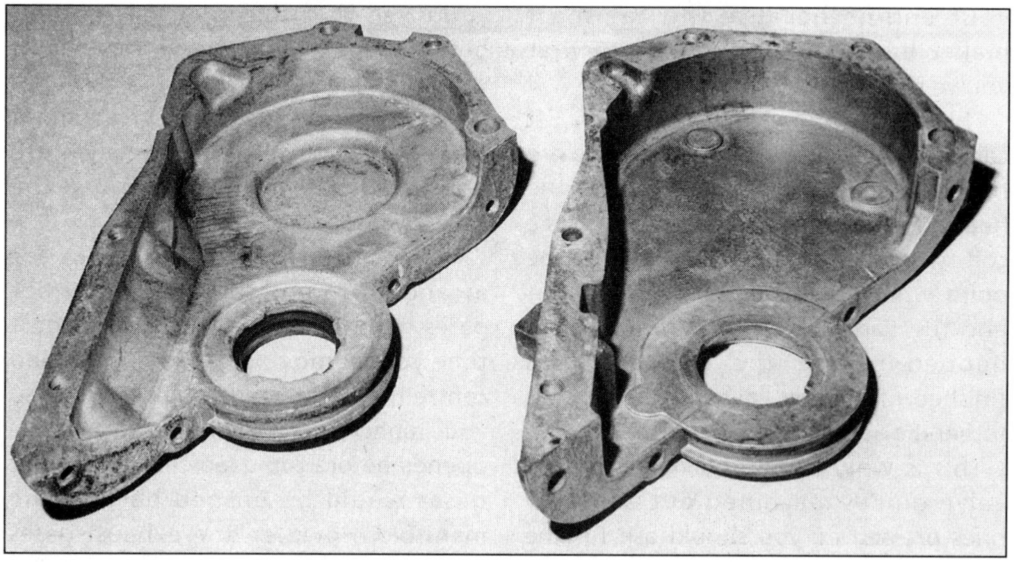

◀ *The timing cover on a Crossflow in a Fiesta (left) is shallower than that on an inline Crossflow, and a normal vernier chainwheel will not fit in.*

▲ *Some tuning firms make special vernier chainwheels for Fiestas, using a series of holes and a peg, like this one from Vulcan Engineering.*

▼ *When you fit a camshaft with extra lift the valve cut-outs in the top of the piston may have to be increased.*

the slight play which develops in each link causes stretch, but on a Crossflow the tension side of the chain isn't very long – barely five links. If each link developed 0.005 in play, which would be really worn out, the camshaft timing would be retarded by less than three quarters of a degree. By that mileage, factors other than stretch in the timing chain would have had a far greater effect on reducing performance.

However, many tuners who take Crossflows up to full race power, and I'm talking here of 135 plus bhp, prefer to be on the safe side and use a duplex, or double-row, chain. So, on the basis that, if race tuners use it, then it must be a good thing, many home tuners also use a duplex chain. If you want to, then with one exception there's no reason, except the cost, why you shouldn't. A duplex chain and its chain wheels will fit all in-line mounted Crossflows without modification. The exception is the transverse mounted Crossflow fitted to pre-1983 Fiestas. These engines have a thinner timing chain cover than the in-line mounted engines and there isn't enough room for a duplex chain.

When the camshaft grinder is reprofiling a shaft, or even machining a performance cam from a new blank, it isn't always possible to maintain the exact angular positions of the cams relative to the dowel which positions the chain sprocket. Even with a standard camshaft, production machining tolerances mean that this angular relationship is not always exactly as specified, but an engine in standard production tune is tolerant to small variations in camshaft timing.

A tuned engine with a high-performance camshaft is far less tolerant. If you want the best from your tuned Crossflow you should always use an adjustable sprocket, sometimes called a vernier sprocket, and set the timing with a timing disc and a dial test indicator (DTI). Once again, Crossflows in Fiestas are awkward because of their thinner timing case cover. A normal vernier sprocket won't fit in, but several tuning firms – Vulcan Engineering, for example – make special vernier timing sprockets for Fiestas, which use a series of small holes and a dowel.

Timing is always carried out on No. 1 cylinder, on the inlet cam. The first step is to set No.1 piston accurately to top dead centre. Because of production tolerances this may or may not be exactly indicated by the TDC mark. The

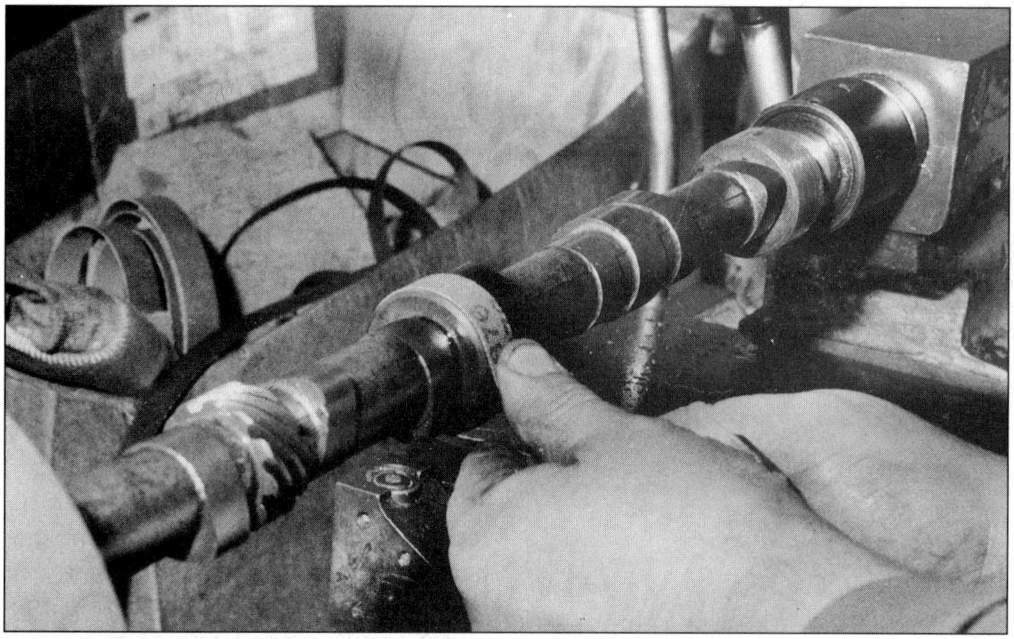

◀ *If you are reusing your old camshaft, clean the journals thoroughly with fine emery cloth.*

only accurate way to set it is with the DTI. Position to the TDC mark on the engine first. General manuals on timing will tell you to position the DTI with its probe through the sparking plug hole and resting on the top of the piston crown. On a Crossflow rebuild you won't be doing the job with the head on, so put the DTI on the top face of the block with its probe on No. 1 piston. Rock the crankshaft back and forth until you get accurate TDC indicated by a change of direction of the DTI needle.

Next, fit the timing disc on the nose of the crankshaft and fix a piece of stiff wire under any convenient bolt head at the front to act as pointer. Bend the wire till it is in line with the 0° mark on the timing disc. You now have the engine and timing disc set, so look in the information supplied with the camshaft to find the full-lift position of the inlet cams. On a popular fast-road performance camshaft for the Crossflow this is 108° so, in this case, you would turn the crankshaft clockwise looking from the front – that is, in the normal direction of rotation – until the 108° mark on the timing disc is opposite the wire pointer.

This is the crankshaft position at which the inlet valve on No. 1 cylinder should be at full lift. Slacken the setscrews in the adjustable chainwheel, put a pushrod in the inlet valve cam follower for No. 1 cylinder and put the DTI probe on the other end of the push rod. Rock the camshaft by turning the centre part of the adjustable chainwheel without moving the chain until No. 1 inlet cam follower is at full lift. Once again, this will be indicated by a change of direction of the DTI needle. Then lock up the ring of setscrews round the adjustable sprocket.

In the event that you are doing this job with the head on, put the DTI probe on the spring cap of No. 1 valve.

When the engine is built you can, as a final check, rotate the engine and check on the timing disc the points where the valves lift off their seats and return. Compare these figures with the timing figures supplied by the camshaft maker.

Chapter 11

The crankshaft and flywheel

If the crankshaft shows signs of ridging on the journals it's completely unfit for further service, even on an engine which is just being overhauled, let alone one which is being rebuilt with tuning in mind. The only recourse is to have all the journals reground, and use oversize bearing shells.

Even if the surface looks perfect, the journals must be checked for ovality and taper. It is also advisable to check the centre main journal for run-out relative to the front and rear mains, the big-end journals for length, for deviation from parallel to the mains and for alignment to the mains. The thrust washer faces should also be checked for being at right angles to the centre bearing journal. If the equipment is available, it isn't a waste of time to recheck these after the crankshaft comes back from grinding, particularly if it has been Tuftrided.

Tuftriding, as mentioned earlier, is a hardening process involving bringing the crankshaft up to a high temperature in a chemical bath and then letting it cool slowly. It increases considerably the shaft's resistance to wear and, it is claimed, also increases its strength and reduces its liability to surface damage which could lead to fractures. It isn't a hideously expensive process and, though I am reserved about it's being worthwhile for a standard rebuild, or even a mild tune, it is worthwhile if you are taking your tuning much higher up the scale. It has one disadvantage – it tends to distort the crank slightly, so accurate checking of all the crankshaft parameters, not just bearing shell fit, is essential after this process.

Ovality and taper on the journals can be checked with a micrometer or vernier calliper gauge, the type with a dial read-out being easier to use than the older type with only a sliding scale. The other checks are best carried out using a dial test indicator (DTI) with the crankshaft set up between centres or, much less convenient, on V-blocks on a surface plate.

If you are overhauling the engine, as distinct from rebuilding, you may want to get away without having the shaft reground. If the journals look in very good condition – no scores or ridges and the old bearing shells have a nice even wear across them and have not worn through the white metal coating at all – all you have to check is ovality and taper.

How much ovality and taper is permissible with new bearing shells? Sorry, but I can't give you an answer. Taper isn't usually a wear problem unless the con rods are bent – which leaves us with ovality. All I can say is that I've known engines with up to 0.003 in ovality on the big-end journals – eight times the permitted ovality on a reground journal – run without

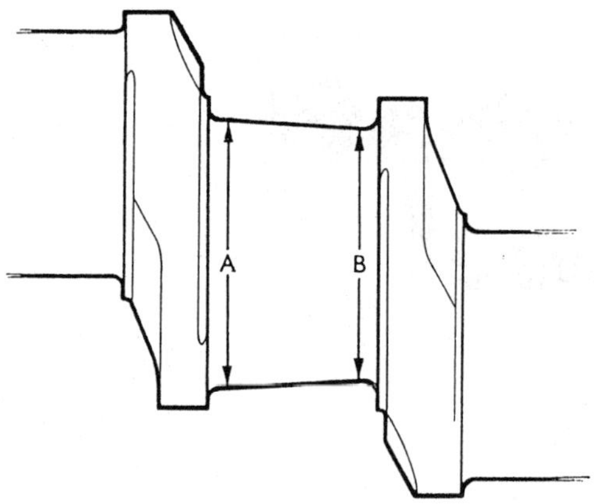

A>B

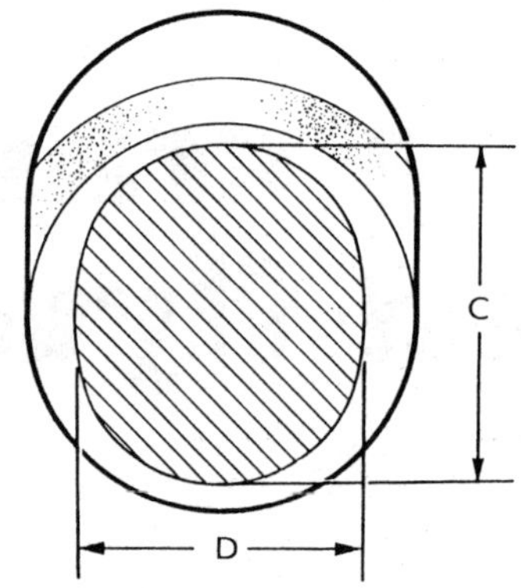

C>D

Taper on a journal is when dimension A is greater than dimension B. Ovality is when dimension C is greater than dimension D. On a reground shaft, taper should not be more than 0.0005 in and ovality not more than 0.0004 in.

problems or knocking for thousands of miles with new lead-indium shells and an oil pump with a high pressure spring kit. On the other hand, I have known engines with this amount of ovality, or even slightly less, knock their big-end shells out in a couple of thousand miles when they have been tuned to a medium fast road performance. Much depends on the way the car is driven.

My own view is that you can get away with 0.001 in ovality with a mildly tuned engine, provided you don't thrash it, but we can't accept any responsibility at all if you decide not to go for a regrind and then knock out the bearings. For peace of mind, have the shaft reground.

When you have it reground, it pays to go to a respected tuning firm rather than a local engineering works, unless the local firm has a good reputation with engine tuners. You may pay a little more, but they will carry out all the checks necessary on a reground shaft and – most important – they will have sophisticated grinding machinery which will grind all the big-end journals to the same throw.

This is a tremendous help on the higher stages of tune if you have the assembly 'decked'. Decking means having all the pistons the same height below the face of the block when they are at top dead centre. When you assemble an engine with Hepolite pistons you'll find that, unless the face of the block has had to be quite heavily refaced, the tops of the pistons will be between 0.010 in and 0.015 in below the top of the block. For a road-going engine, even in fast road tune, this variation is nothing to worry about. If, on the other hand, you are going for top racing power, and money isn't a prime consideration, have them brought even.

Be guided by the tuning firm, because what the deck height should be depends on the make and type of piston, and on the type of head gasket you are going to use. If it varies on a dry-run build, the firm will be able to advise you whether the tops of the pistons can be machined to bring them to equal height or whether it's necessary to skim the top face of the block.

On a reground shaft, ovality on both main and big-end journals should not be greater than 0.0004 in (0.010mm). Taper should not be greater than 0.0005 in (0.013mm). If a DTI is used, these measurements should be total indicator reading (TIR).

The centre main bearing run-out relative to the front and rear journals should not be greater than 0.002 in (0.050mm) TIR. The big-end journals should be parallel to the mains within

0.001 in (0.025mm) and a centre-line through each one should lie within 0.005 in (0.013mm) either side of a centre-line through No. 1 big-end and the front main journal.

The length of the centre main bearing journal can be increased by up to 0.020 in (0.51mm) provided an equal amount is machined from both faces and oversize thrust washers are fitted. Check with the person doing the grinding that the appropriate thrust washers are available. Most tuning firms stock them in oversizes from 0.002 in (0.051mm) up to 0.015 in (0.381mm). The length of the big-end journals should not be increased by more than 0.010 in (0.25mm).

It is important when a crankshaft is being ground that the correct fillet radii at the ends of the journals are maintained. If they are too large, the edges of the bearing shells will pinch and run hot. If they are too small, they could set up stress concentrations which could lead to a crankshaft fracture at high revs if the shaft reaches its natural resonating frequency.

The centre and rear main bearings have a double radii. The inner radius must be maintained at 0.070 in (1.78mm) and the outer radius at between 0.100 in and 0.110 in (2.54mm to 2.79mm). All the other main and big end journal radii should be between 0.080 in and 0.094 in (2.030mm to 3.290mm). The radii should be smooth and you should not be able to see any chatter marks on them. These could also lead to stress concentrations.

Both before the shaft goes for grinding, and after it comes back, the oilways should be squirted through with white spirit using a syringe. Squirt from the oil hole in each main journal and watch the white spirit as it comes out of the big-end oil holes. Continue squirting till it comes out absolutely clean, than blow through with compressed air.

Except on engines for use with automatic gearboxes, which don't use a spigot bearing, there is a needle roller spigot bearing in the back of the crankshaft to take the clutch shaft of the gearbox.

It's a very tight press fit in the end of the shaft, and you will certainly have difficulty in getting the old bearing out

▼ *These are the fillet dimensions to which a crankshaft grinder should work.*

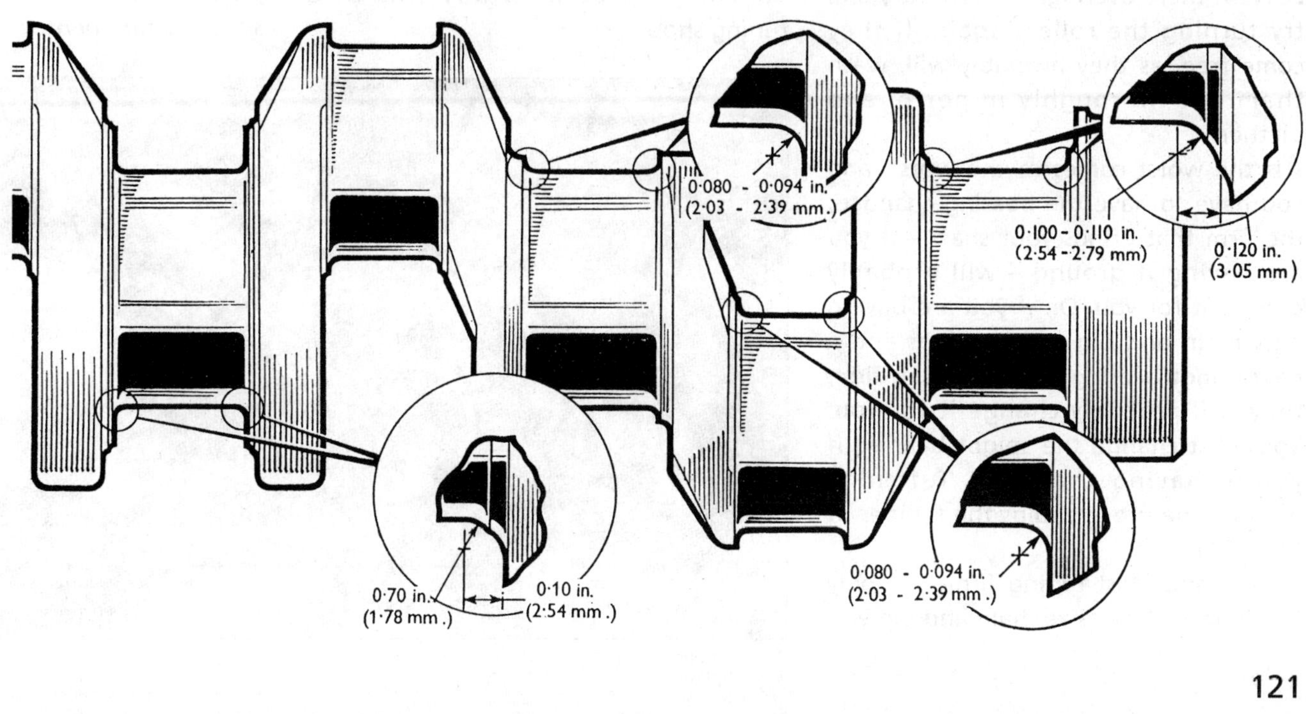

in a home garage. Ford has a special tool for extracting the old bearing, and others are available from specialist tool firms, but they are expensive and, unfortunately, don't have a reputation for lasting very long. The claws which hook behind the bearing have to be so fine that they tend to round off. If the bearing has to be changed, some people grind it out with a small riffler stone, but the material is so hard that you need a really high speed tool – 20,000 to 30,000 rpm – and even then it takes a long time. You'll never do it with an ordinary electric drill. Other people break the bearing up with a small cold chisel.

Fortunately, the bearing needs changing on only a relatively small number of engines. As a rule of thumb, if you put your finger in it and you can rotate the rollers freely, it's OK. Just give it a good wash out and oil. What you sometimes find is that the rollers are stuck so you can't turn them. Provided they haven't got flats on them, try standing the crankshaft on its nose (well supported) and fill the bearing cavity with WD40 or a similar releasing agent. Leave it there overnight if necessary and try turning the rollers again. If they come free, as they probably will, wash them out thoroughly in petrol and oil them.

If the worst comes to the worst, and you have to have the bearing changed, the firm that grinds your shaft – if you are having it ground – will probably change it for you. Or, if you are buying new main and big-end bearings and a new spigot bearing from a tuning firm, they will probably change it for you. You must change the spigot bearing if you're having the shaft Tuftrided because the process ruins the temper of the rollers.

Ford, and most tuning firms, list only three sizes of oversize main and big-end bearing shells, plus 0.010 in, plus 0.020 in and plus 0.030 in. Before you buy your new bearing shells, however, check whether the outside diameter is standard or oversize. New shells are available in both standard and oversize outside diameters.

As explained in Chapter 12, it's possible that the internal diameter of the main bearing housings has been bored out to 0.015 in (0.381mm) oversize to correct misalignment or ovality, even on new engines. The old bearing shells should be marked on their backs if the outside diameter is oversize, but this is not infallible. Either take the old set with you for the bearing supplier to measure, or torque down the main bearing caps to the block and measure the internal diameter of the bearing housings. The standard housing bore should be between 2.271 in and 2.272 in (57.683mm to 57.696mm). An oversize bore should be between 2.286 in and 2.287 in (58.064mm to 58.077mm). You can measure this with an internal micrometer or vernier calliper, but there is a go, no-go plate gauge available. You may be able to buy this at a tuning shop.

▼ *Taper and ovality can be checked with a micrometer or vernier callipers. Checking other parameters needs special equipment.*

▲ *Crankshaft balancing is essential in a tuned engine, but it is more convenient to have the clutch balanced separately on a slave flywheel.*

▶ *The standard flywheel on a 1600 Crossflow is quite a weighty affair.*

▶ *It can, often with advantage, be lightened by machining metal off the inside of the rim.*

▶ *This is taking flywheel lightening to the limit for a racing engine. The tickover would be very lumpy.*

Big-end shell bearings are not available with oversize outside diameters because the housing in the con rod should not be oversize. Shells for reground crankshafts are available in a range of sizes from 0.002 in to 0.060 in oversize on internal diameter.

Some people may tell you, or you may read, that it's advisable to have the big-end journals cross drilled. The theory behind this is that when the shaft is revolving at high speed, centrifugal action tends to fling the oil away from the bearings. With cross drilling, the old oilways in the journals are blanked off and new oilways are drilled at an angle to lessen the centrifugal effect. On some engines, Ford's V6 is an example, there can be a definite benefit but, in my view, it isn't necessary to have this done on a Crossflow shaft, even for competition work. Certainly, on a road going engine it's an unnecessary expense.

Some people have a fetish about lightening the flywheel. Ford has used a number of weights of flywheel with different capacity Crossflows, ranging from 16.75 lb (7.60 kg) to 26 lb (11.8 kg), the lighter flywheels usually being on the GT versions. A light flywheel certainly gives you an engine with a quicker response to the throttle and better acceleration but, if you have it too light, you won't get a smooth tickover.

Tuning for power with a high-performance camshaft and possibly higher compression ratio also gives you a less smooth tickover. You have to trade one off against the other. Take the advice of the camshaft maker. He will want you to get the best from his product in terms of response and acceleration, but he won't want you complaining to everyone that fitting his make and model camshaft gives you an engine that ticks over like a camel with hiccups.

If you have the flywheel lightened, make sure that whoever does it knows what they're doing. Probably the best person to entrust with the job is the firm that balances the crankshaft and flywheel. They will know where it is safe to take off metal, and they will also know about the importance of leaving a good radius at the base of any machining. Any general engineering shop will probably lighten it for you, but if they do it wrongly there's a chance that it might fracture at high revs. It's an outside chance, but outside chances have been known to happen and the consequences of a flywheel disintegrating at high speed are frightening.

What is essential on a tuned engine is to have the crankshaft and flywheel balanced. This will give you a much smoother running engine. Some people like to have the shaft, flywheel and clutch balanced as a complete assembly, but there is a drawback here. Yes, if you're aiming for high revs it's advisable to have the clutch balanced, but the advantage of having it balanced together with the crank and flywheel doesn't outweigh the disadvantage if you have to change the clutch.

It's better to have the shaft and flywheel balanced together and the clutch balanced on a separate balanced slave flywheel. That way, if you have to change a clutch later, you can have the new one balanced by itself and know that your engine is very close to perfect balance instead of wondering whether or not its worthwhile stripping down the engine to balance the new clutch with your shaft and flywheel.

Nose Pulley

It isn't part of the crankshaft, but this is as good a place as any to talk about the nose pulley. Ford's standard nose pulley is a pressed metal affair made in two halves. It works fine and lasts well on an engine in standard tune. But, when you tune to higher power it has a distressing habit of the two halves parting company. Much better is a pulley machined from solid which you can buy at most tuning shops. Usually you'll find it's a smaller diameter than the Ford one, but this is no bad thing.

It's still quite large enough to drive the alternator fast enough to charge at tickover but, being smaller, it drives the alternator and water pump slower than the Ford pulley. The difference isn't great, but it's enough to stop the water pump rotor cavitating at high revs and running in a pocket of air with static coolant round it. With an engine in standard tune, the revs don't get fast enough so the problem doesn't arise.

▲ *A smaller steel crankshaft pulley stands high revs better and helps to stop the water pump cavitating.*

▼ *A machined steel crankshaft pulley compared with the standard Ford pressed one.*

Chapter 12

The block and crankcase

A Crossflow block and crankcase from a road-going car has to be very badly worn indeed, or damaged, to be scrap. The only reasons I know for rejecting it are accident or frost damage, a piston ring circlip coming adrift and allowing the gudgeon pin to come out and score the bore or, not very often, a block where the cylinders have been overbored to their maximum and have then worn to the point where they would need reboring again. Even in these cases, it's possible to have liners pressed in, though it's a lot cheaper to get another block because there are so many of them about.

You should check the top face of the block with a straight edge for warping, but it isn't a common fault. Usually it needs only to be cleaned off thoroughly. Even if you have to have it refaced, the amount of metal which will have to be taken off will be so small that it's unlikely to affect things like piston deck height.

What, then, does need attention on the block? Obviously, the cylinder bores wear. When you strip and measure your cylinders you may be puzzled, even upset, to find that the cylinder bores are already 0.015 in (0.38mm) oversize on the standard 81mm bore size, even though you know that the car has its original engine which has never been rebored.

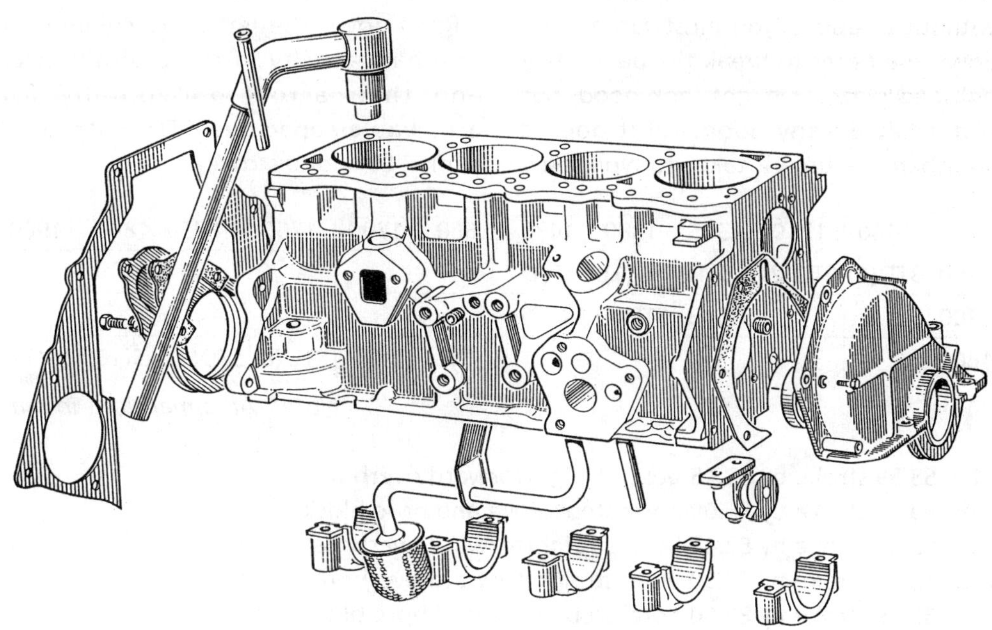

▶ *The Crossflow block and crankcase is robust and seldom gives problems.*

There's nothing wrong. For reasons best known to itself, Ford sometimes built new Crossflows with 81.38mm bores instead of the standard 81mm.

If you're just overhauling an engine to give it a new lease of life with maybe a mild tuning, rather than rebuilding with more serious tuning in mind, you've got to make your mind up whether or not you're going to have the cylinders rebored and fit new pistons. Here, I'm going to chicken out and say that the decision must be up to you. I can't lay down any hard and fast figures about how much cylinder wear is permissible.

What I will do is repeat that if, after cleaning the carbon off the unworn top edge of the bore with wet and dry paper you can feel a ridge when you drag your fingernail up it, think seriously about reboring. If you can't feel a ridge, and only see a discoloration, you could get a lot more miles out of the engine with a set of new rings before a rebore becomes necessary. It will take mild tuning, too, because that doesn't accelerate bore wear all that much, it's harder on bearings than bores.

If you decide you can get away without a rebore, you must run a hone down the bores to break the beautifully polished glaze. It might look good, but it doesn't do any good. All it does is lengthen the time it takes for your new rings to bed in and accelerates bore wear because it can't retain any oil and the rings run dry.

You can buy an inexpensive cylinder bore hone that will fit in your electric drill at most good tool shops. The stones are spring loaded so one hone covers a range of bore sizes. Lubricate it well with paraffin, run your drill at a medium speed and run the hone up and down the bores till they have an even grey finish. Then wash of all traces of the grey carborundum.

If you decide you're going for a rebore you have to decide to what oversize you're going to have the bores taken. On a rebuild to standard tune, or if you have only a moderate increase in power in mind, you need only have them rebored to the next size up that will take out the wear. If you're going for extra power it may be worthwhile taking the bores up further to gain on capacity. If you want to go for even more capacity than straightforward overboring or even sleeving can give, you can combine them with a long-stroke crank. The following table, from Oselli Engineering at Oxford, shows what can be achieved.

Read down the left hand column for original capacity in cubic centimetres and then across horizontally for achievable capacities. STD = standard stroke, LS = long stroke.

	1140	1210	1345	1380	1430	1660	1700	1760	1825	1840	1900
1100	STD(1)	STD(2)	LS(1)	LS(1)	LS(1)						
1300			STD(3)	STD(4)	STD(5)						
1600						STD(6)	STD(7)	STD(8)	STD(9)	LS(10)	LS(11)

Notes. All dimensions in mm.

1 53.29 stroke by 82.45 bore: straightforward overbore
2 53.29 stroke by 85.00 bore: step sleeve and bore block
3 62.99 stroke by 82.45 bore: straightforward overbore
4 62.99 stroke by 83.50 bore: straightforward overbore
5 62.99 stroke by 85.00 bore: step sleeve and bore block

6 77.62 stroke by 82.45 bore: straightforward overbore
7 77.62 stroke by 83.50 bore: straightforward overbore
8 77.62 stroke by 85.00 bore: step sleeve and bore block
9 84.00 stroke by 83.50 bore: overbore to siamese block only
10 84.00 stroke by 83.50 bore: straightforward overbore with special crank
11 84.00 stroke by 85.00 bore: step sleeve and bore block with special crank

The most popular fast road overbore conversions for the 1300 and 1600 engines are the ones which take the bore up to 82.45mm, a 1.45mm overbore, commonly called a sixty thou overbore, and 83.50mm, a 2.5mm overbore, commonly called a ninety thou overbore, and stay with the standard stroke. This takes the 1300 up to 1345cc or 1380cc and the 1600 up to 1660cc or 1700cc nominal.

The 1100 isn't so often tuned, as most people prefer to change to a 1300 or 1600 engine as a cheaper way of getting more power, but those who need to stay with 1100cc for insurance or other purposes, and still want to tune their engine, usually go for a straightforward 1.45mm overbore, commonly called a sixty thou overbore, and the standard stroke. Sleeving and special cranks to get greater capacity is more usually done only for competition purposes.

Like the small-end bushes, camshaft bushes on the Crossflow seldom wear very much. If they look all right and are not scored from grit in old oil that hasn't been changed for years, it's not usually necessary to replace them for a normal fast road engine except, possibly, the front one which has to take the sideways pull of the timing gear. Changing them at home isn't easy without special tools so, if they do need changing, let the tuning firm do it. If there is any camshaft bearing change to be done, check before you get the new bearings that the old ones are not oversize on outside diameter. There are two sizes, standard and plus 0.020 in (plus 0.508mm).

At the bottom end of the crankcase there's only one area where there might be some work to do and that's the main bearing housings. Problems here are much less prevalent than with the older pre-711M engines which had curved main bearing caps. After long mileages these used to stretch so that the housing was oval. The standard cure was to let the caps down and then line bore and ream them back to standard. Just occasionally it's still necessary with the heavier, more square 711M bearing caps so, if you're going for higher power it's worthwhile checking them. As mentioned in the previous chapter, the standard vertical internal diameter with the cap torqued down should be between 2.271 in and 2.272 in (57.683mm to 57.696mm). For top power for racing, steel main bearing caps are available from tuning firms, but they aren't necessary for road use.

▼ *You can have the block of a 1600 Crossflow bored out to 0.090 in oversize to bring the capacity up to 1700cc. Smaller engines can be similarly uprated.*

▲ *After boring, the block is honed to exact size.*

◀ *Honing should leave the surface looking lightly scratched.*

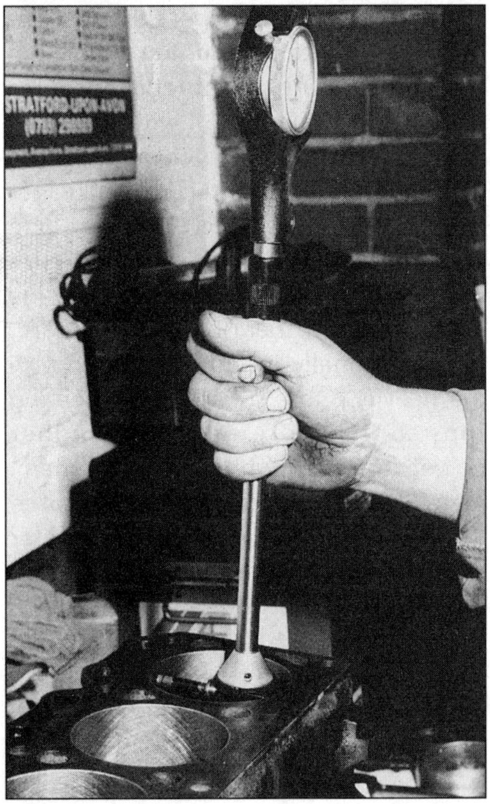

◀◀ *Checking a bore dimension with a dial test indicator.*

◀ *When you're cleaning a block, syringe and blow air through the oilways.*

▶ *Camshaft bearings seldom need replacing provided the oilways are syringed and blown clear.*

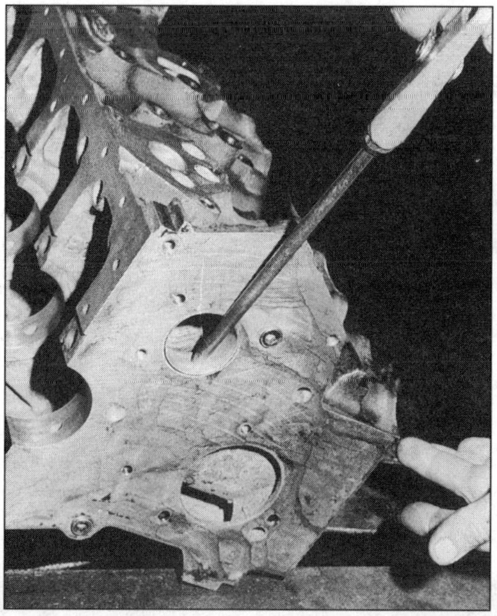

▶▶ *Check that the locating tags of the shell bearings sit perfectly in the bearing housings.*

▲ *For racing power some tuners fit Wills rings – extra fire resistant rings – into recesses machined round each cylinder.*

▶ *Some tuned Crossflows use the older type of breather outlet in conjuction with a remote oil separator. The petrol pump mounting has been blanked off here as an electric pump was being used.*

Chapter 13

Pistons and connecting rods

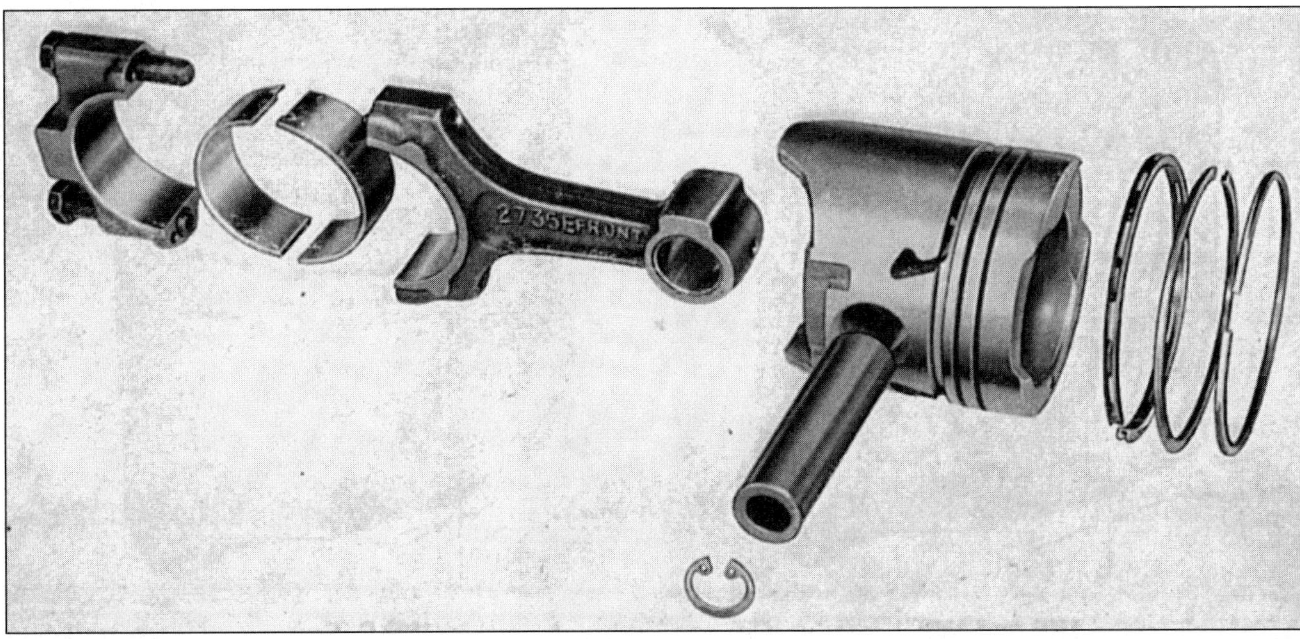

▲ *Piston and con rod assembly.*

Apart from a few earlier engines which had a very small combustion chamber in the head, the Crossflow uses a Heron combustion chamber with a flat face to the cylinder head and the chamber in the bowl of the piston. It has advantages, and most of the racing Crossflows producing very high horsepowers use the Heron design. But there are also drawbacks and, over the years, a number of tuning firms have tried to overcome these and have modified the flat-face heads by machining them to produce a combustion chamber.

The first drawback is that the Heron design means using a heavier piston and, when you're going for high rpm, the heavier Heron piston puts more stress into the con rod and crankshaft than a lighter, flat-topped piston. Some tuners reckon that, without this extra weight, the rods and shaft could safely be taken up a further 400 to 500 rpm before the stress started to get near danger level.

The second is that, because most of the extra weight is concentrated above the gudgeon pin, the piston tends to rock in the bore, particularly when passing top dead centre. This gives a more rapid bore wear in the area where you want it least, the top quarter of the bore where the sealing between the rings and the cylinder wall takes the full blast of the high combustion pressure. For normal tuning and fast road work, the degree of extra wear needn't worry you too much. Far more increase in wear is caused by neglecting oil changes so that the rings become clogged and stick in their grooves.

The third drawback is the greater area of a Heron piston which is exposed to the combustion heat and its rather complex cross-section. The greater area allows heat to be conducted more quickly away from the combustion area which raises the temperature of the pistons and means that less of the heat in the burning fuel is available to convert to horsepower. If you remember your physics lessons at school, the main law of energy states that energy can neither be created nor destroyed, only changed in form. The only source of energy we have to drive our engine is the fuel. The more of this energy we can convert to drive the crankshaft round and the less we lose as heat to the cooling system, the better.

▼ *Most Crossflow pistons have valve cut-outs like the one on the right. Pistons from Escort Mk I 1300 GT engines (left) did not.*

▶ *Pistons are graded 2.134 in down from the top of the crown at right angles to the gudgeon pin.*

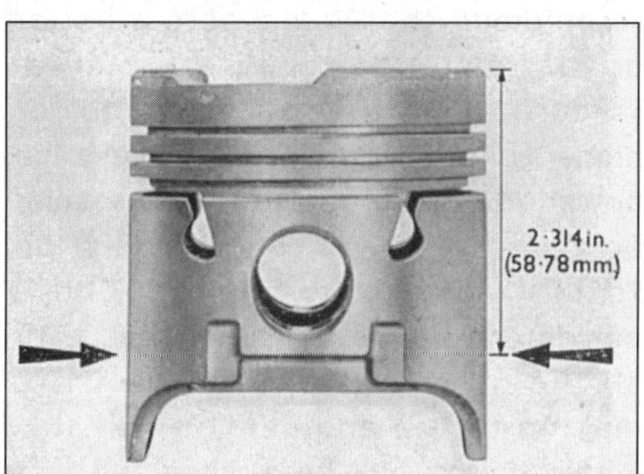

The cross-section of a Heron piston, particularly when it has cut-outs machined for the valves, is quite complex with fairly abrupt changes in cross-section area. This means that the temperature of the piston is much higher in some places than in others which, in turn, means that its expansion is far less even that that of a conventional flat-topped piston.

Indeed, when you go for high power with a Heron piston, whoever is doing the tuning has to keep a very close eye on temperatures. The higher temperature associated with a weak mixture, for example, can tip the scales to the point where the pistons start to burn.

This is one of the big dangers that home tuners can encounter if they follow the advice of some saloon-bar pundits and use pistons from a 1300cc engine in a 1600cc engine to push up the compression ratio. It will do it, because the bowl in a 1300 piston is smaller than that in a 1600 piston, but, as some people have found to their cost, the pistons can have quite a short life.

If you are seeking high compression ratios, then probably, a chambered head from one of the tuning specialists is your best route with a 1300cc or 1600cc engine. With some high-lift camshafts, though, you might still need cut-outs in the piston crown to clear the valves.

With a 1600cc engine bored out plus 0.090 in, to give you a nominal 1700cc, there's barely enough metal in the head to allow for a combustion chamber to give you a compression ratio within reason so you would need a piston with at least part of the chamber in the crown, which tends to negate the advantages of a lighter flat-topped piston. The same applies if you want to keep the compression ratio down to the level where you can happily use normal pump fuel, either four-star or super-unleaded.

All in all, the general opinion of tuners is that, although on a flow bench the figures for a chambered head may be higher, there isn't all that much to choose when it comes to peak power output – but the chambered head and flat-topped piston may give you slightly better mid-range torque figures.

If you are planning for short sprint racing, or specializing in hill climbs with short, twisty courses, the faster response of an engine with the lighter flat-topped pistons could give you an advantage. For circuit racing, where the engine is at high rpm for longer periods, or for general fast road driving, you are probably better off saving your money to use on other parts of the car and staying with the usual Crossflow flat-face head.

Valve Clearances

Changing the subject to pistons and valve clearances: when you buy your performance camshaft with its higher lift than the standard Ford item, the maker should be able to tell you (all other things being equal) whether or not you will need to deepen the cut-outs in the piston crown to give sufficient valve clearance.

However, all other things seldom are equal. You may have had the face of the head or the block, or both, skimmed to offset warping, or you may be using rocker arms with a higher than standard ratio. All or any of these, plus normal machining tolerances, can affect the amount by which the valve protrudes down towards the piston on full lift, and thus the clearance between the valve and the top of the piston. You may also have to use a head with larger valves to give you greater gas flow, in which case the old piston cut-outs will no longer be large enough even if they're deep enough.

You can calculate the valve clearance, but the only safe way when you deviate from standard is to check your calculations with a 'dry run' engine build. As described in Chapter 10, spread a thin layer of Plasticine in the piston's valve cut-outs and turn the engine over a couple of times. Then you can measure the thickness of the Plasticine where it has been squashed down by the valves. Anything less than a clearance of 0.060 in (1.52mm) at the bottom of the cut-outs is unsafe. Even with its five main bearings, the Crossflow's crankshaft can flex and the con rods and pistons expand with heat. With too small a clearance there is a danger of the valve heads kissing the top of the pistons. You should also allow a clearance of at least 0.040 in (1.02mm) between the sides of the valve head and the sides of the cut-out.

Just like sharp edges to the valve heads, sharp edges to the piston cut-outs can glow red hot and be a possible source of pre-ignition. Play it safe and have the corner at the bottom of the cut-out and the rim at the top finished to a slight radius.

Most tuning firms will steer you away from standard Ford pistons, even if you are having the cylinders rebored to one of Ford's standard oversizes. They will recommend that you choose a better quality piston – Hepolite is quite a favourite – or, if you need to go for higher compression ratios to achieve the power you want, pistons with a smaller bowl in the head. You can go from Ford's standard 9.0 to 1 up to 10.0 to 1, but do check with the tuner that, with all the other modifications you are making to the engine, you will stay within the piston's heat range and that you can still use standard pump fuel with the optimum ignition advance. Unless you are tuning for racing, there's no need to use very expensive forged pistons. Cast pistons are quite suitable for all fast road work.

You should, however, have the set of pistons checked to see that the height above the gudgeon pin bore is the same for all four. Remember that, if you have to have metal removed to bring them to the same height, this will affect the compression ratio. You should also have them balanced. The machinist will pick the lightest one and then machine the insides of the heavier ones to make them all the same.

Connecting Rods

The four connecting rods should also be the same weight. If you have the choice, try to pick four that are the same weight without alteration but, if any have to be altered, entrust them only to a machinist who knows what he's doing and will machine metal from the small boss at the bottom of the cap. If they are lightened wrongly, there's a danger of setting up stress concentrations which, at high revs, might cause the rod to fail.

Some people will tell you it is worthwhile going to the expense of having the standard Ford rods shot peened and polished to relieve stresses. They look beautiful when they're done but, with a road-going engine, it's debatable whether or not the expense is worth the gain. To do any good, the shot peening has to be carried out at a high pressure and, in nine cases out of ten, this bends the rod, so you have the extra expense of having it straightened. It would be a different matter if Crossflow rods had a history of breaking, but they don't.

I have heard of only two reasons for a con rod parting company with the crank. Either the rod had been damaged by careless fitting or taking apart and there was a blemish which started a crack because of stress concentration, or one of the big-end bolts failed. Providing your rods are not damaged, the chances of one of them breaking at the revs you will achieve with a road-going engine are so small as to be discounted. With regard to big-end bolts, get a set of high-tensile competition bolts from a Ford Rally Sport main dealer. If they're torqued up properly they won't give way.

What you must do, however, is to check each rod carefully to see that there are no dents or blemishes perhaps caused by a previous careless assembler who has marked them by gripping them in unprotected vice jaws, hit the caps with a metal faced hammer or blemished them in any other way. Any blemish of this sort is a prime place for stress concentration and could possibly lead to cracking and failure at high power and high revs.

You should also have them checked to make sure they are straight – a surprising number of used con rods aren't. If, when you took the pistons out, there were signs of heavy wear marking on one side above the gudgeon pin and corresponding heavy marks on the other side below the gudgeon pin, it is an almost sure sign

▼ *Con rods from different size engines are different lengths. Check the part numbers if you are making up a set.*

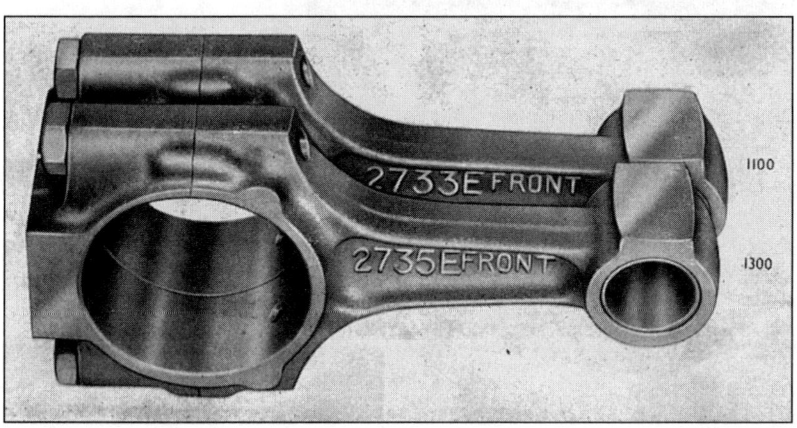

that the con rod is bent. Also have the rods checked to make sure that the big-end and small-end bores are perfectly parallel and that the big-end where the bearing shells fit is truly circular. If they are oval, ask the advice of a tuning firm. They may say that you can have the cap let down and the housing remachined, but generally it is far cheaper to find another rod.

If you are building an engine from a 'basket' of parts, or if you are swapping con rods to find four of equal weight, be careful that you don't muddle up the con rods from 1100, 1300 and 1600cc engines. Because the bores on all the Crossflows are the same, the crankshaft throws and con rod lengths are different. There's no real excuse for mixing them up because the part number is on the side, together with the word 'Front'. The part number of the 1100 rod is 2733E, the 1300 is 2735E and the 1600 is 2737E.

▲ *Pistons are balanced to find the lightest one, then the others are machined down inside to match the weight.*

Also, if you are not using the original four con rods, remember that they have the numbers 1 to 4 stamped on the camshaft side of the rod and cap at the big end to identify which bore they are used in. If you use rods from different

◀◀ *Checking a con rod for alignment.*

◀ *After the alignment check, con rods should be weighed on double scales and machined if necessary to balance them.*

engines you will end up with two or more rods with the same number. This will confuse not only you but also the next person who strips the engine. Obliterate the old marking and stamp a new number to correspond to the bore in which you intend to use the rod.

The small-end bushes in Crossflow con rods don't very often wear. Ford quotes tolerance figures on diameter, but the easiest way to test them is to try pushing a new gudgeon pin through them. If it fits snugly without rock and pushes through with easy thumb pressure, the bush is fit for further service. If the pins falls through very loosely, or rocks, the rod needs a new bush.

If you are stripping and overhauling an engine that doesn't need reboring, and using the pistons again, you will want to use new rings, which means that the ring grooves in the pistons will need to be cleaned out. Clean the piston first by soaking it in white spirit to loosen the carbon, and then cleaning it with a soft brass brush. Don't, please, use a steel wire brush.

If the pistons showed markings that indicated a bent con rod I would think very hard before using them again, particularly if you have tuning for extra power in mind. You will almost certainly get piston slap. Go for a new set of pistons and, for the relatively small extra cost, have the block rebored and honed so that you start again from square one.

You can buy a special tool to clean out the ring grooves or, if you're careful, you can do it with the old rings. If you snap an old piston ring, the sharp break makes an excellent scraper to clean out the grooves. Don't go so deep that you take a lot of metal off the bottom of the grooves, though a tiny amount does no harm because the ring depends for its sealing on a small amount of gas passing behind it and expanding it out towards the piston wall. Too much bottom clearance, however, leads to gas blow-by. Keep the broken ring accurately in line with the groove to avoid taking metal off the sides which would give your new rings a sloppy fit. Make sure that the oil drain holes and expansion grooves behind the rings are clear.

As I've said before, however slight the bore wear in an engine, there will be a slight ridge at the top of the bores and, even though this might show up as a discoloration rather than something you can feel with your finger nail, the sharp edge of a new ring will find it. Make sure that you get a set of rings with a 'ridge dodger' top ring. This has a tiny cut-out round the edge of the top surface to clear the ridge in the bore. Make sure you fit it the right way up.

It is a matter of choice whether or not you go for standard oil control rings or special ones, such as Cords, for a used engine that hasn't been rebored. If you are not reboring we would always advise the special rings. There used to be a lot of talk that these rings hastened bore wear, and they were nicknamed 'rebore while you drive' but, though there may have been some justification for this with some makes of rings years ago, modern special oil control rings won't wear the bores any faster than plain oil control rings. You may or may not have to have the oil ring groove machined to accept the special oil control rings. This used always to be the case, but with most modern brands it is no longer necessary. The supplier will tell you.

Chapter 14

The lubrication system

The lubrication system on the Crossflow is straightforward and generally very reliable. Two types of oil pump have been used on the Crossflow a sliding vane type and a bi-rotor type. They're interchangeable, and you can identify them from the outside because the bi-rotor type has four recesses in the end-plate whereas the end-plate of the vane type is flat. The bi-rotor is more efficient than the vane, so if you have a vane type it's worthwhile changing it.

Both types of pump have a full flow filter built in to them and, on the earlier pumps this is the type where you have to take a canister housing off the pump and renew the element inside. Later pumps have the much more convenient throw-away cartridge-type element which you just screw off and on. Again, it's worth changing to the later type.

Oil is drawn up from the strainer in the sump to the oil pump which is driven by a skew gear from the

◀ *Two types of oil pump were fitted to the Crossflow. The earlier vane type can be identified by its plain end-cover.*

▶ *The main identification feature on the later rotor type pump is an end-cover with four recesses.*

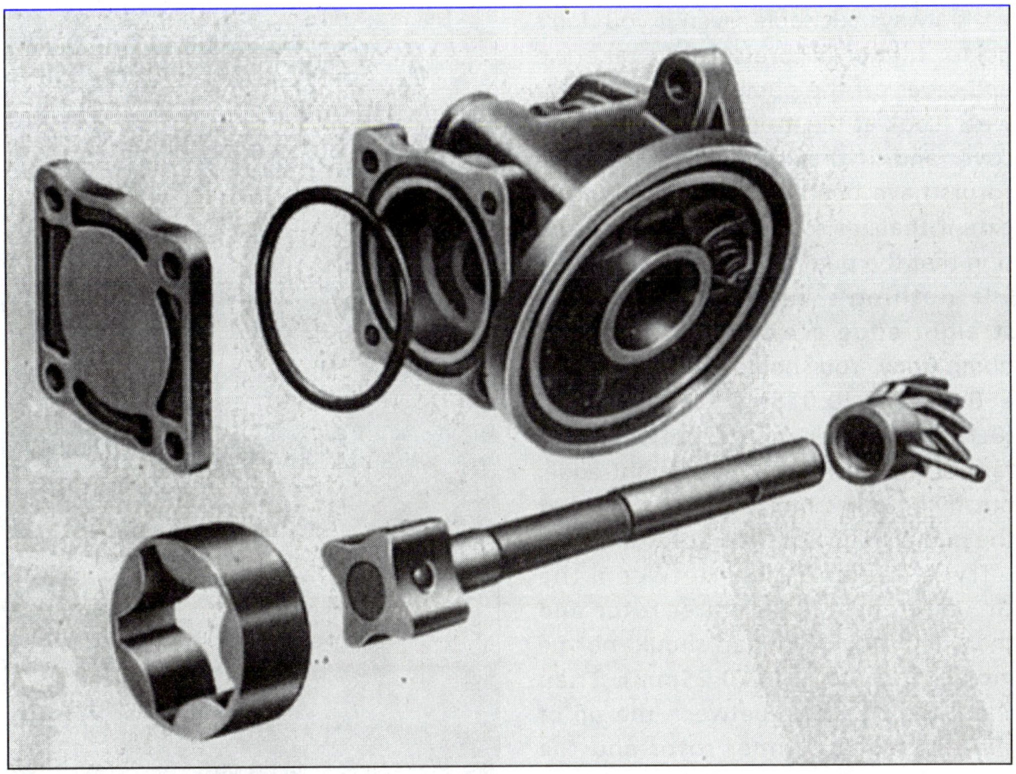

camshaft. There's a plunger-type relief valve incorporated in the pump, and oil which passes through this is returned to the sump via a pipe which ends under the oil level to avoid frothing.

Oil under pressure goes through the filter to a short gallery on the right hand side of the engine, and this is tapped to take the sensor for the oil pressure warning light or the adaptor for the oil pressure gauge if you have one. There's also a cross-drilling through the crankcase to a longer main oil gallery which runs along the left-hand side of the engine.

From this gallery there are five drillings to take oil to the main bearings. There's a chamfer on the centre main to feed oil to the crankshaft thrust bearings, and drillings in the front, centre and rear main journals feed oil via the crankshaft to the big-end bearings. There's a small drilling in the web of each con rod to squirt a jet of oil to the non-thrust side of the cylinder bores and the small-end bushes each time the shaft revolves.

The camshaft bushes are also fed via drillings from the front, centre and rear mains, and the timing chain is lubricated by a small jet of oil from a metered hole in the front drilling.

There's a flat machined on the front camshaft journal which controls the feed to the rocker shaft, so you get a jet of oil under pressure to the rocker shaft each time the engine revolves. This oil goes through drillings in the cylinder block and head to the front rocker pedestal and then through the hollow shaft to lubricate the rockers. There are drillings in each rocker to lubricate the valve stems and pushrod ends, and this oil drains down to lubricate the tappets and cams on its way back to the sump.

It's important to keep this circulation in mind when you're cleaning all the parts after you've dismantled the engine so you make sure all the passages are clear.

It's always advisable, even if you don't go for higher oil pressure, to take the end-cover off the pump and check it for wear. Look at the inside face of the end-cover and, if that shows signs that the rotors have been rubbing on it to the extent that it's scored, it's a pretty sure sign that the pump is well worn.

If nothing's wrong there, put a straight-edge across the face of the pump body. You should be able to push a 0.003 in (0.075mm) or 0.005 in (0.13mm) feeler gauge between the face of the rotors and the straight-edge, but if the gap's much bigger than that the pump will be down on pressure.

Try a feeler down between the circumference of the outer rotor and the pump body. The gap should not be more than 0.010 in (0.25mm). Then check the clearance between the tip of the lobes of the inner rotor and the lobes of the outer rotor. This should not be more than 0.006 in (0.15mm).

If these clearances are excessive, it's a pretty sure sign that the rest of the pump is worn and it isn't worth stripping it down to try to renew worn parts. The oil pump is the heart of the engine and pumps its life blood, so it isn't worth taking chances. Fit a new one.

If all the clearances are OK, the pump's fit for further service unless you want to fit a higher than standard pressure. If you do, you can either buy a new high-pressure pump from one of the tuning firms or, if your pump is unworn, you can get a kit with an uprated relief-valve spring and plunger to push up the pressure on your pump. One supplier of these kits is Burton Engineering in Ilford, Essex.

You will see high-pressure plus high-capacity oil pumps listed in tuning catalogues but, unless you're building a racing engine, you don't need the higher capacity, providing you change the oil regularly. So use a high-pressure

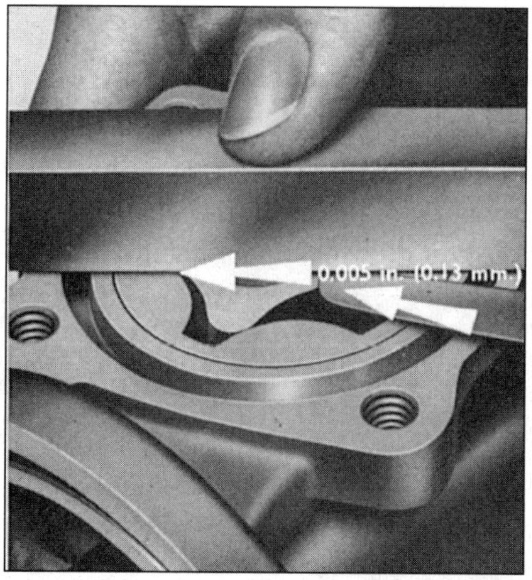

◀ With a straight-edge across the body of a rotor type pump the gap for a feeler gauge should not exceed 0.005 in.

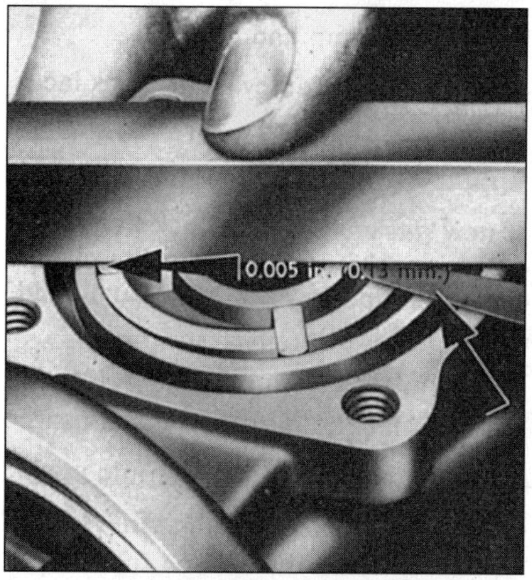

◀ The same maximum gap of 0.005 in applies to a straight-edge across the face of the vane type pump.

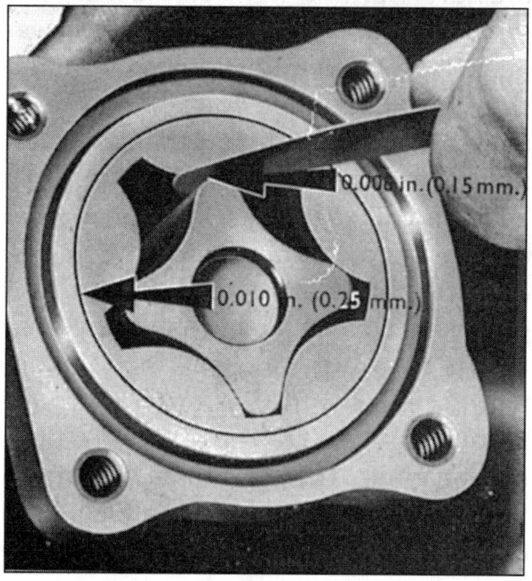

◀ Checking the clearances of the rotors. The gap between the outer rotor and the body should not be more than 0.010 in. Between the tip of the inner rotor and the lobe of the outer one it should not be more than 0.006 in.

▶ *If you're tuning for higher power, an uprated, higher pressure oil pump is a good investment.*

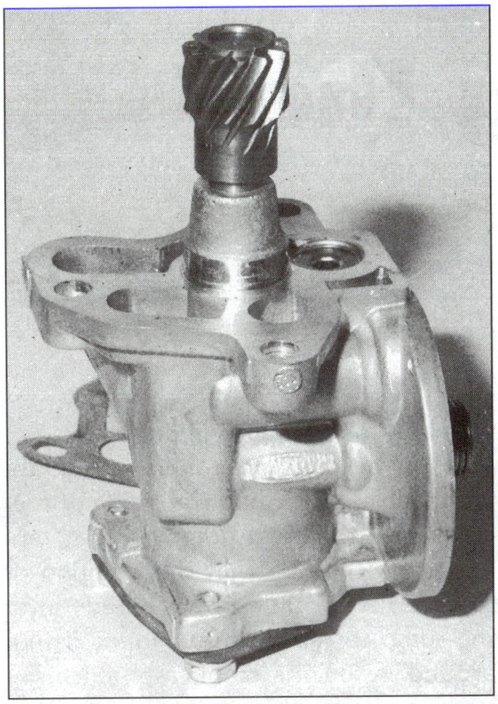

▼ *If your engine has this earlier type of filter, change the oil pump for a later one which takes the more convenient throw-away canisters.*

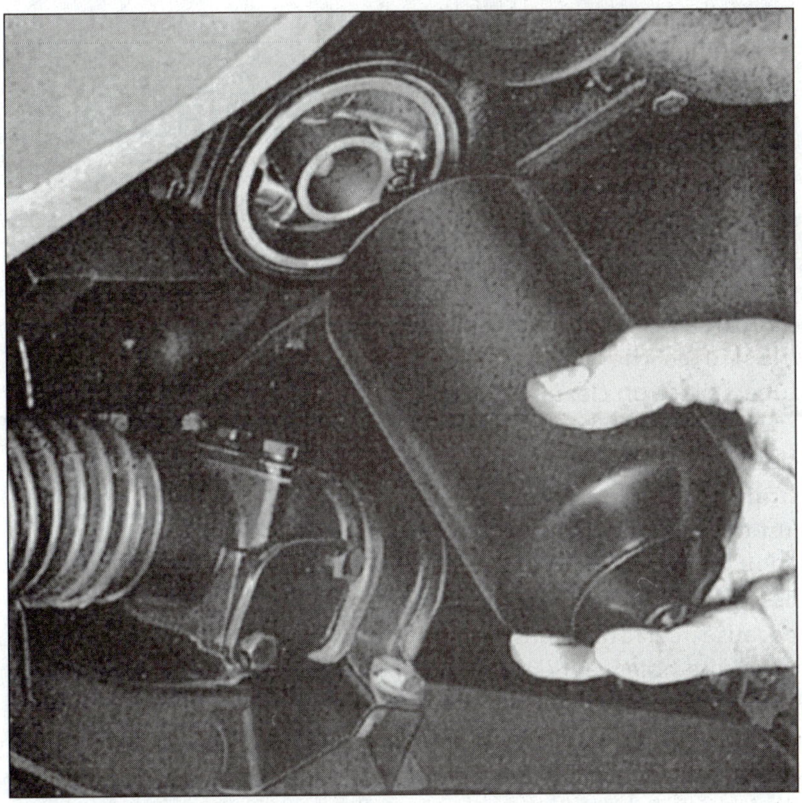

pump and save the extra money to spend on other parts.

You will also find kits to fit oil coolers and for dry-sump lubrication. This is a system where the oil isn't kept in the bottom of the sump, it's kept in a separate tank and a very shallow sump is fitted. The oil pump is, in effect, two pumps in one. The pressure side of the pump delivers oil just the same way as the standard pump then, when the oil has drained back down into the sump, it's picked up through a pipe by the scavenge side of the pump and sent through an oil cooler to the reservoir tank.

This is a system strictly for racing and is usually used on short circuit racing where the almost constant centrifugal force of cornering pulls the oil in a normal sump over to one side so that the normal oil pick-up pipe could start sucking air. It's very expensive, complicated to fit and very costly, and it definitely isn't needed on a road-going car.

A plain oil cooler is another matter if you are tuning for competition work with sustained high speeds where the oil could possibly get overheated. Even then, modern oil technology has produced oils that will withstand far greater temperatures before they deteriorate than oils of even a few years ago. Before you decide to spend out on an oil cooler, have a word with a couple of major oil companies to get their views. My guess is they will tell you that a top-quality modern oil with regular oil changes is more efficient, and cheaper, than fitting an oil cooler for any work except, possibly, racing.

Chapter 15

Rebuilding the engine

Cleanliness, said the Victorians, is next to godliness. Whether you believe that or not, there's no arguing that cleanliness, methodical working and constant checking are the secrets of success when it comes to reassembling an engine. The higher you go in power tuning, the more important constant checking and dry-run testing become, so clean the bench, clean your tools and clear the floor before you start.

You might think that all the checking I am going to advise is unnecessary and that it's a theoretical counsel of perfection. If you buy new parts, or have parts reconditioned, surely they ought to fit together properly. Certainly you won't find this sort of checking on commercially rebuilt engines in standard tune, but that's not what you're doing. You're spending a lot of money, time and effort to build an engine that's better than standard. Time spent checking is well spent, even if you don't find a single fault.

I usually start with the block, but that's a matter of habit as much as anything else. When you get your block back from reboring, or pick up your block again after honing the cylinder bores, it may look clean but I doubt very much if it is. Have a really close look round inside the crankcase to make sure there aren't any cakes of old dried up oil and dirt lurking in nooks and crannies. If there are, get them out and then wipe out the inside with a clean rag dipped in petrol.

I know of some race tuners who go to the trouble of painting the inside of the crankcase, though I don't go that far. I've heard theories that this painting is to help the oil slide down quickly into the sump to reduce the amount flying around inside the crankcase which, in turn, reduces the resistance to the crankshaft turning, but I've always regarded that as armchair theorizing, and taken it with a large pinch of salt.

One race tuner to whom I told that story laughed and said that he painted the inside so that when the engine was stripped down for checking, which was quite often, it was easier to clean out. I don't think it's necessary to go to those lengths on a fast road engine, but it illustrates the importance that top tuners place on cleanliness.

I will take it that you attended to cleaning out the water jacket when you stripped the block, but it's a good idea just to check that all the loose muck is out.

Put in the new core plugs. Do it now, before you get involved in other things, because it's fatally easy to forget one when the hole is hidden by other things hanging on the engine.

You can buy a special compound for sealing the core plugs at most Ford main dealers, and though there are excellent proprietary sealing and jointing compounds on the market I usually use the Ford one because it's made for the

▶▶ *Before you put the new core plugs in smear the special paste round the hole and on the plug.*

▶ *Put the plugs in position making sure they're square.*

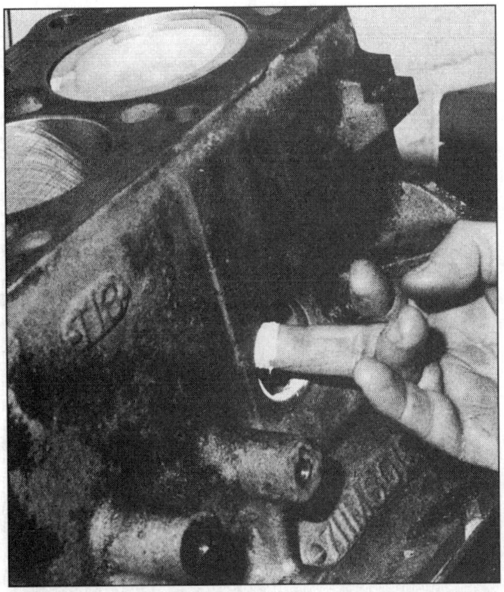

▶ *Drive them in flush with the block using a mandrel – an old socket is ideal.*

▶ *Wipe off the excess paste and throw the rag away.*

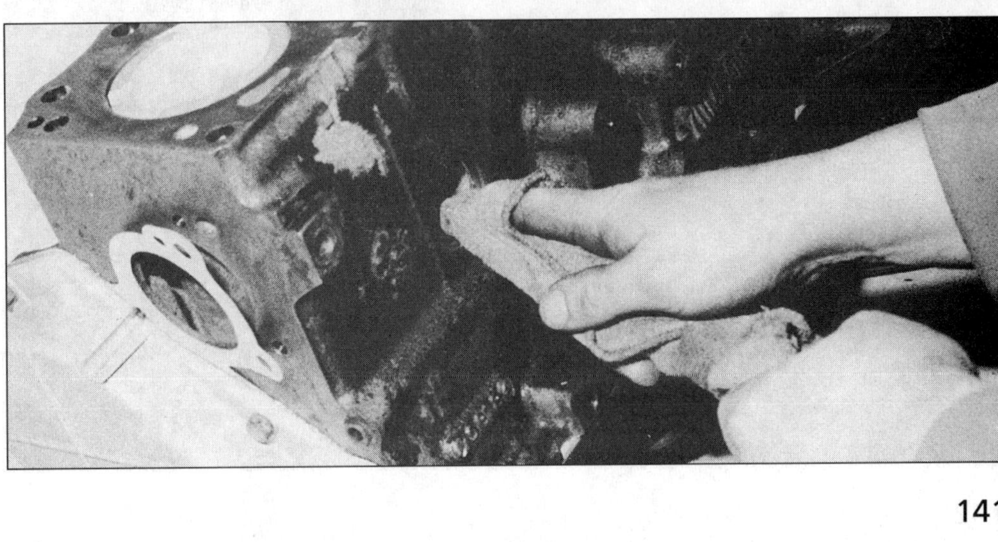

◀ *Some race tuners secure the core plugs with cap screws, but I haven't found it necessary on Crossflows.*

◀◀ *After reboring, wipe round the inside of the bores with a rag dipped in transmission fluid to clean out the last of the honing dust. Throw this rag away as well.*

◀ *When you gap piston rings, push them down in the bore with a piston to make sure they're square.*

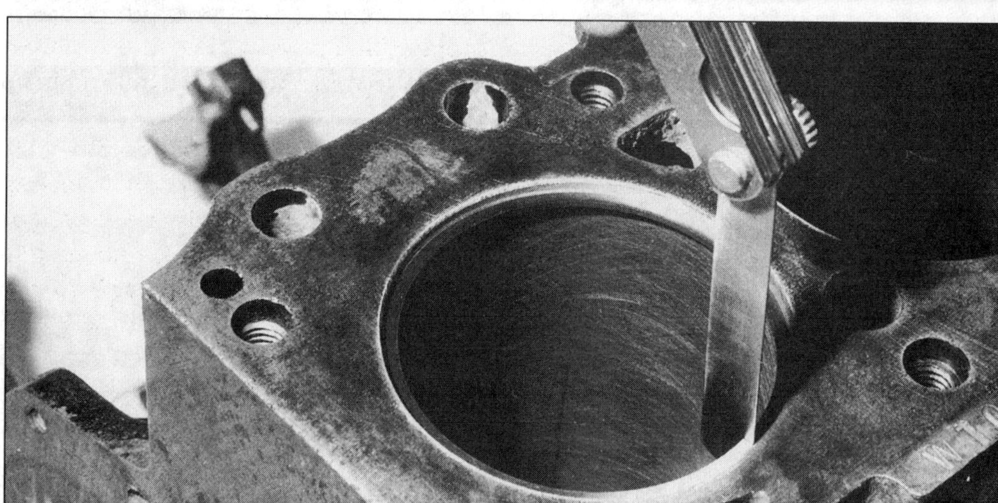

◀ *Check the end gaps with a feeler and adjust if necessary.*

job and I've never had a core plug leak.

Coat with compound the inside face of the hole in the block, and the sides of the new core plug, and tap the plug into place until the edge is just level with the face of the hole. Don't drive it in by hammering on the rim, as that's as good a way as any to get it in crooked, and you run a real risk of distorting the rim so it doesn't go in cleanly. Use a reasonably well-fitting mandrel of some sort inside the 'lid' of the core plug. Some people use a length of round section wood, but I usually use an old socket which fits inside the core plug, and put it on a short extension bar to give me something by which to hold it.

Some race tuners always lock the core plugs in position; some never do. I have seen racing engines on which core plugs have popped out, usually at the back. On some engines – the Pinto is one – I'll go along with the theory that at certain high revs there is a vibration in the block which can loosen them. I haven't heard of it happening on a Crossflow, which is a more dumpy and stiffer block, but if you want to play safe and lock the plugs, the way to do it is to drill and tap three equally spaced holes round the core plug just far enough away not to distort the edge of the hole. Twice the diameter of the tapped hole away from the edge of the core plug is a minimum. Then put in three setscrews and washers (see illustration, opposite, top left), coated with Loctite or a similar thread locking compound. Most of the tuners I know who do this use Allen-headed machine capscrews, but hexagon headed ones work just as well.

Wipe the oozed-out compound from round the core plugs and throw that piece of rag away so that you don't get compound on any of the bearing surfaces. Then get a clean piece of rag, moisten it lightly with automatic transmission fluid and wipe round the insides of the cylinder bores. Transmission fluid is the best thing I know for picking up all traces of dust and carborundum that inevitably stay behind in the bores when they are washed out after honing. Continue cleaning till there's no sign of grey dust left on the rag, and throw that one away too.

With the bores thoroughly clean, the next job is to check the fit of the pistons in the bores – something I touched on earlier in Chapter 13. The person who bored and honed your block should have had the pistons to hand and he should have checked with a piston-pull gauge when he did the honing. A piston-pull gauge is like a small tubular spring balance with a thin feeler gauge attached to it.

The piston, without its rings, is pushed into the bore with the feeler gauge on one side, and the pull to extract the feeler is measured. Each bore should have been honed to a particular piston, and the piston boxes should have been numbered.

Different tuners work to different clearances. Some like a close fit, 0.0009 in to 0.0010 in (0.023mm to 0.043mm). An engine with these clearances will need running-in. Others hone to a wider clearance, 0.0020 in to 0.0025 in (0.048mm to 0.064mm) on the basis that few customers run their engines in properly and a wider clearance is better than a piston skirt which picks up in the bore.

These two clearances represent a pull of between 7 and 11 lb on a spring balance with a 0.5 in wide feeler gauge. If you're working to metric figures, the width is 12.7mm. These are fairly standard widths for long piston-clearance feeler blades. If you want to check for yourself, a good engineering tool stores ought to be able to order them for you complete with the spring gauge.

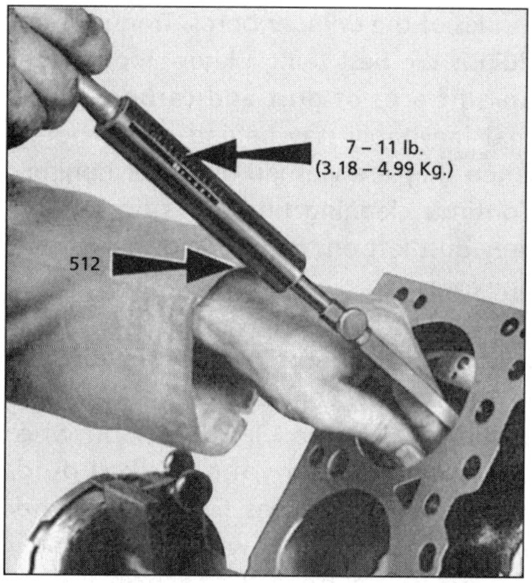

◀ *The best way to check piston fit is with a pull gauge.*

▼ *Check the side clearance of the rings with a feeler gauge.*

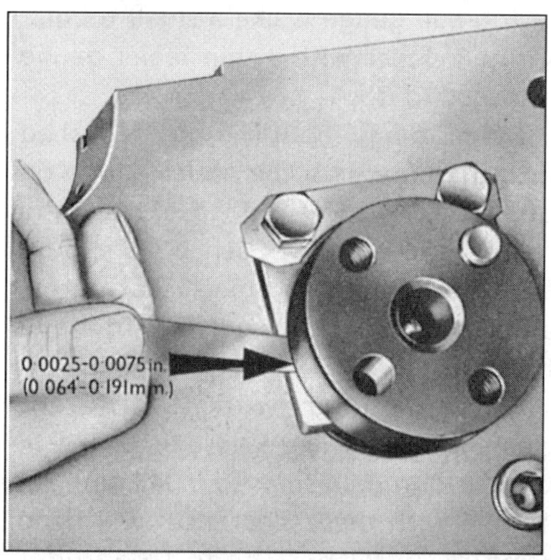

◀ *Before you lock up the camshaft bolts, check the end float.*

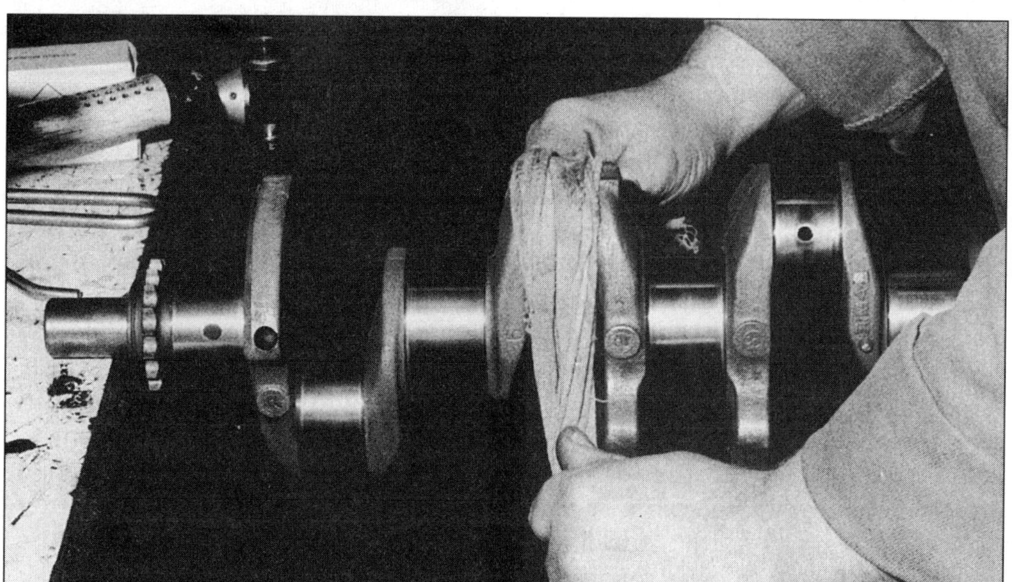

◀ *Give the crankshaft a thorough clean and syringe out the oilways.*

You may not feel it's worthwhile going to the expense of buying a piston pull gauge but, at the very least, you should take the rings off the pistons – if they aren't already off – and check each one down a bore. Don't worry about getting them the right way up, it doesn't matter. The clearance you're feeling for is the skirt clearance, not the clearance at the rings which will be larger.

If you stand the block on its face on a piece of soft rag so the air doesn't rush out of the bottom, then with both the bore and the piston clean and dry, the piston should drop down the bore evenly and steadily against the damping of the air pressure. If it whips down the bore and hits the bottom with a thud, it's an indication that the bore's on the large side. If it requires a hard pressure to help it down, particularly if it sticks, the bore's probably on the tight side. I say 'indication' and 'probably' because this is a very unscientific way of measuring piston to bore clearance but, if you've got any doubt, ask whoever rebored the engine what clearances they worked to. It's better, of course, to ask before you hand the job out. If you get a blank look and non-committal reply like: 'We work to standard overbore sizes', try somewhere else.

The average clearances I have given here all go by the board if you're using high-expansion alloy pistons. These start off with a much higher clearance and whoever does the reboring must work to the piston maker's specifications.

You've got another job to do before you can put the pistons away in their boxes, and that is to check the ring gaps and side clearances. Piston rings must have a gap between their ends when they're put in an engine because they expand with the heat. If the gap is too small, the ends of the ring will butt together and the ring will go tight in the cylinder and possibly break or seize.

To check the gap, put each ring in the cylinder in which it's going to work and push it down gently with an upside-down piston to get it square in the bore. Then measure the gap between the ends with a feeler gauge. Unless the piston maker specifies something different, as he may well do if you go for something exotic, a good working figure for the ring gap is 0.009 in to 0.014 in (0.23mm to 0.36mm) I generally work to an easy sliding 0.012 in because it's a handy size feeler gauge. If the gap is too large, you can waste power because of gas blow-by which will also lead to earlier deposits of carbon in the ring grooves. A gap which is too small, however, can be disastrous.

Too small a gap is unlikely in these days of computer controlled machinery, but it can happen. If it is too small, you've got to widen it by filing a small amount off the end of the ring. This is something to be done very carefully and gently because it's only too easy to break the ring. It should be held in a small instrument maker's vice with smooth jaws, not the big bench vice where the jaws may have scars on them and will almost certainly have a deep cross-hatched pattern on the jaw faces to stop things slipping. Don't use fibre or plastic 'work protectors' in the jaws. Something as delicate as a piston ring is likely to slip round in them.

Hold the ring in the vice with about a quarter of an inch of the end showing and take the end down slightly with a very fine file. A coarse file will snap the ring off. Keep the file square and don't saw backwards and forwards with it. A file cuts on the forward stroke, so lift it off the ring each time you bring it back. Try it frequently in the cylinder bore to make sure you don't overdo it.

The next clearance to check is the side clearance of the rings in the piston.

These are checked with a feeler gauge, and you may find figures quoted in the leaflet you get with the pistons or with the rings if you are re-ringing old pistons. If not, a safe clearance to work to is 0.001 in to 0.003 in (0.025mm to 0.075mm) for compression rings and 0.002 in to 0.003 in (0.050mm to 0.075mm) for standard oil control rings. If you are using new pistons and rings it's most unlikely that the side clearances will be wrong, but there's always the chance – it happened to me once – that the wrong packet of rings was included in the piston box.

If you are fitting special oil control rings, the clearance, if applicable, will be marked on the packet. I say 'if applicable' because some types of special oil control ring use a number of 'wavy rings' (rings with a sideways spring) in the piston groove, so you can't measure any clearance. In this case the width of the groove will be specified on the packet. Some special oil control rings will fit in the standard oil control groove in the piston, but with others you have to have the groove machined wider. Check when you order the rings.

I will deal with fitting the rings to the pistons and pistons to the con rods later but, for now, put the pistons and rings safely away in their boxes and turn your attention back to the block.

Turn the block upside down on the bench on a clean piece of rag and wipe through the cam follower bores with a petrol-moistened rag. Wipe the cam followers and try them in their bores as a dry fit. They shouldn't wobble, but they mustn't be tight. Sometimes, by selective assembly, you can get the best overall fit by trying the followers in several bores.

Some of the followers are awkward to put in unless you've got very slim hands and fingers, and even more awkward to lift out. I use a small magnet on the end of a bar which I find useful for all sorts of mechanical fishing jobs. If you haven't got one, a small rubber sucker on a stick, of the type used for grinding in valves, does the job. If it has previously been used for grinding in valves, make sure all the old grinding paste is cleaned off.

When you've decided on which followers will go in which bore, take them out and keep them in order. The easiest way is the old tried and tested method of putting them through holes in a length of wood or even stiff cardboard. Don't forget to mark one end of the card 'Front'.

The next dry-run check is the camshaft. I say 'dry run', but that isn't strictly true because I like to put just a light smear of oil on the camshaft bearing journals before I push it in place, because I don't like bearings running dry, even on 'dry' runs. Take a little care when you feed the camshaft in, the edges of the cams can be quite sharp and it's easy to catch them on the edges of the bearings in the block and raise burrs.

Put a couple of bolts in the front of the camshaft so you can turn it and check that it revolves freely. I doubt very much that you'll find that it doesn't. I've only ever found one camshaft that was stiff, and that wasn't on a Crossflow – it was on a V8 Chevvy engine. It turned out that one of the camshaft journals had been ground above tolerance. The maker was most apologetic and replaced it with a perfect one, but it shows that people who grind camshafts are human and mistakes can happen. Time spent finding no faults is time well spent.

Now fit the camshaft thrust plate, which also retains it, and check the end float. If you've got one you can check it with a DTI (dial test indicator), but feeler gauge checking is accurate

▶ *To use Plastigage, cut a length of the plastic wire (A) and lay it on the journal.*

▶▶ *After torqueing down the cap, the width of the plastic (A) is measured using a width gauge (B) to find the clearance.*

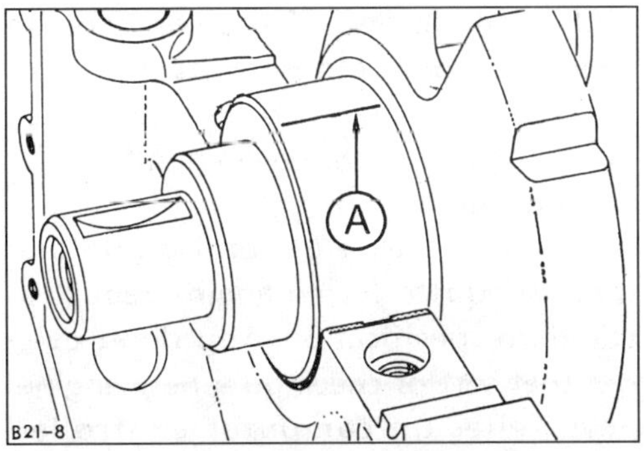

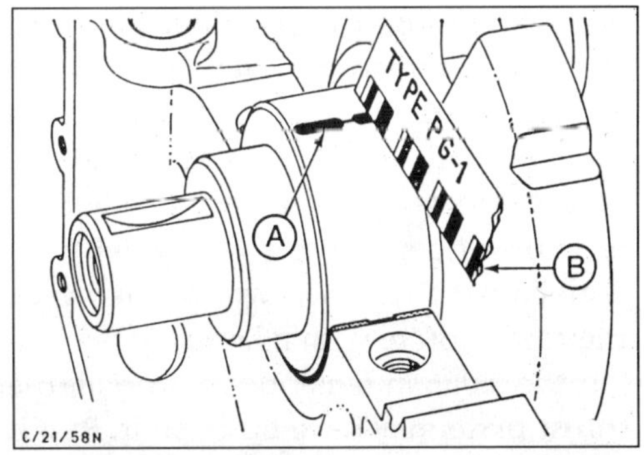

▶ *When you put the camshaft in, be careful not to catch the sharp edges of the cams on the bearings.*

▶ *Use new locking washers and lock the bolts after again checking the end float.*

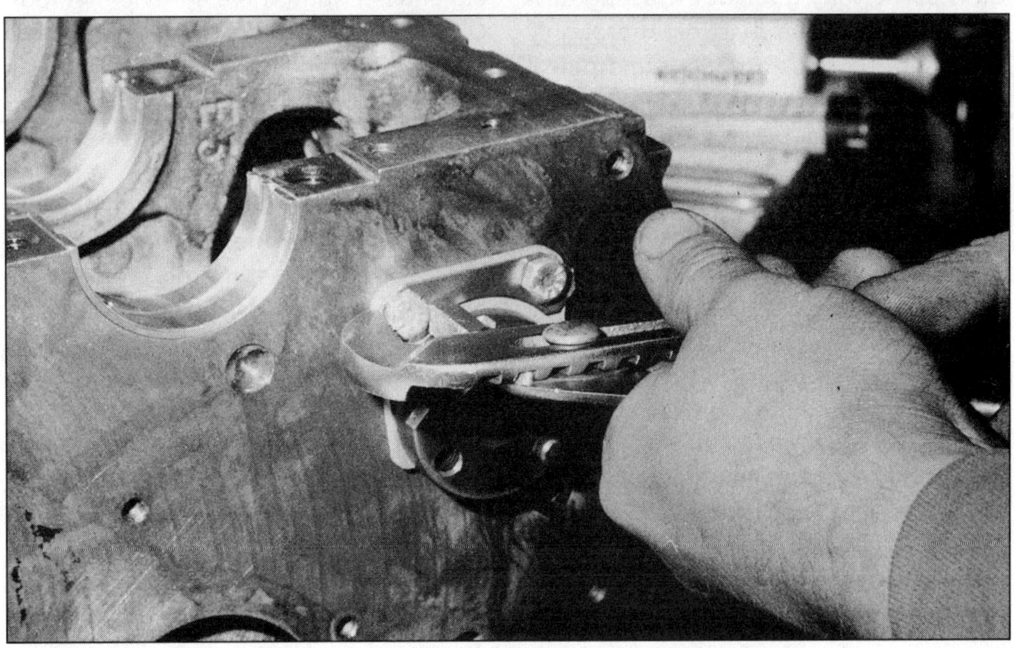

enough because the end float tolerance is fairly generous. Ford gives it as between 0.0025 in and 0.0075 in (0.064mm to 0.191mm). Most sets of Imperial feeler gauges start at 0.005 in, so if that goes in easily, and a 0.010 in feeler doesn't (which will probably be the case), you're within limits.

If the 0.005 in feeler doesn't go in, the thrust plate may have burrs on it. Rub it down slightly using a sheet of emery paper on something perfectly flat. I keep a piece of plate glass in the workshop for this sort of thing. If you have to rub down anything in this way, keep the pressure in the middle of whatever it is you're rubbing down or you'll get it wedge shaped. Do your rubbing down well away from the engine assembly bench, and make sure every last trace of emery dust is cleaned off each time you check it.

Next to be checked is the crankshaft. Your reground shaft should have been cleaned and the oilways blown through before you get it, and it will come back covered with oil to protect the journals. Even though the shaft may have been wrapped up in waxed paper or plastic, oil is particularly good at attracting dust and dirt. Clean the shaft off thoroughly and use a syringe to squirt some petrol through the oilways. There's no danger of diluting the oil because the petrol will have evaporated long before you fit it.

We went through checking the grinding of a crankshaft in Chapter 11, so everything should be OK. Even after initial checking with a micrometer it's never a waste of time to double check by trying the fit of the big-end bearings on their journals and the fit of the shaft on its main bearings.

An old rule of thumb engineer's check for big-end fit is to hold the crankshaft horizontally in the vice – always hold it by one of the webs and use fibre or plastic work protectors in the vice jaws – and assemble one of the con rods on the shaft complete with its bearings, dry. When the rod is held out horizontally it should drop gently under its own weight.

This is fine as a general guide and quite adequate for an engine rebuilt to standard specification. You don't even get that sort of check on a new engine. Ford relies on computer controlled machining and accurate tolerances to get a correct fit.

In the overwhelming majority of cases the fit should be fine, but remember Murphy's Law which states that if something can go wrong then, sooner or later, it will. I prefer to work to Flanagan's Law which states that Murphy was an incurable optimist and that it always goes wrong for you. It's a pretty safe bet that if something goes wrong only once every ten thousand or twenty thousand times, that once is going to be on a part that someone is using to build a high-performance engine. So, if you're building with extra power in mind, now or later, check and double check.

How do you double check the fit of a big-end bearing? There are two methods. One is by using something you seldom see nowadays outside a toolmaker's room or the workshops of some tuners who build high-performance engines, and that's engineer's blue. This is a very fine blue paste, finer than toothpaste. You may not find anyone stocking it locally, but a good engineering tool shop should be able to order it for you. It should come in a tube, so dirt doesn't get in – not in a tin. Wash your hands before you use it, because if there's any grit on your fingers it will stick to the blue and score the bearing shell.

Before you start any bearing checking, make sure that the bearing shells are the correct size for the size you have had the crankshaft reground. The

▶▶ *Wipe clean the bearing housings in the block.*

▶ *Fit the bearing shells with their tags in the locating slots.*

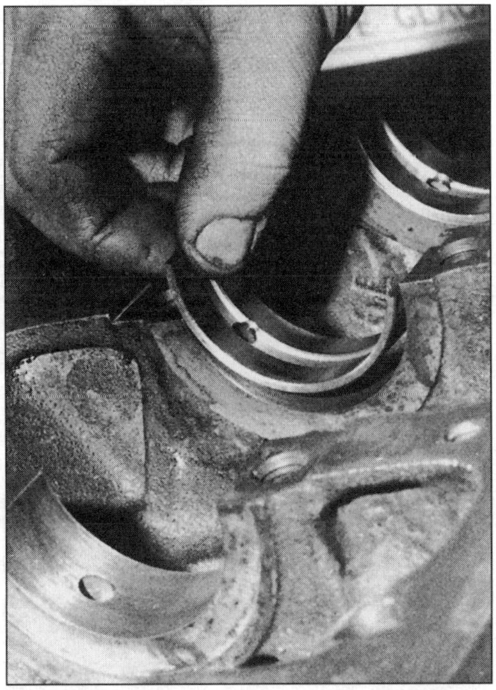

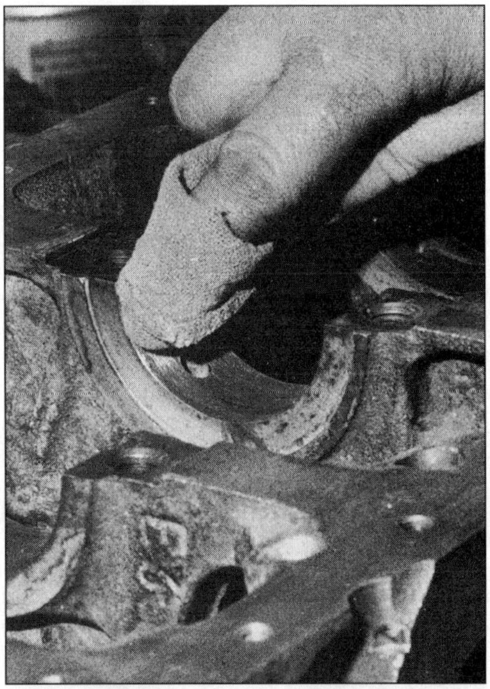

▶ *Then wipe the shells to make sure there's no dirt on them.*

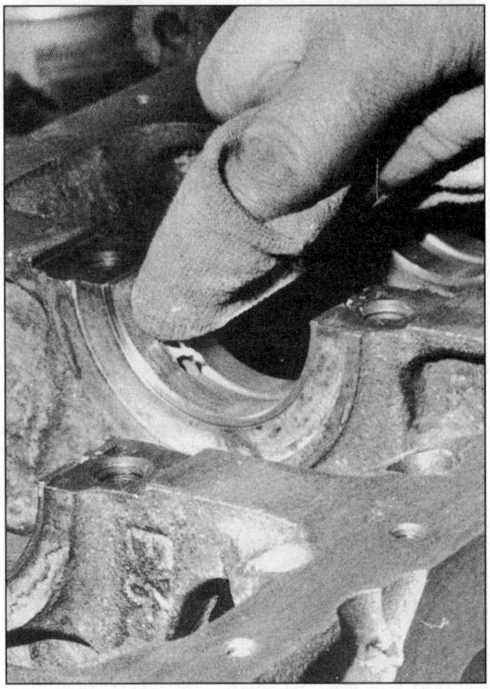

▼ *Squirt plenty of oil in the oilways and put some graphite compound on the shells.*

▲ Lower the crankshaft in place.

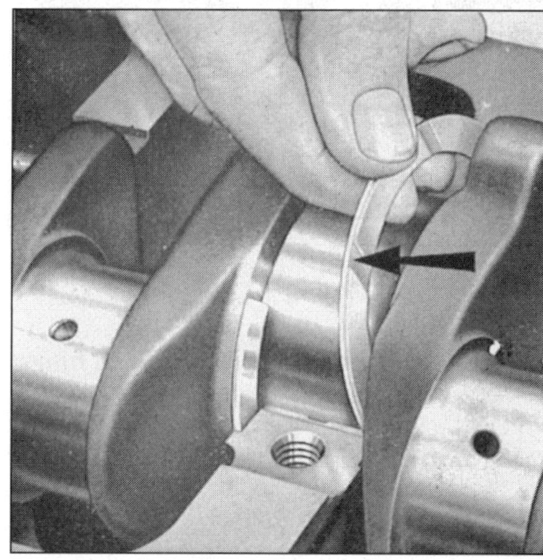

◀ Fit the thrust washers either side of the centre main bearing.

◀ Check the end float of the crankshaft at the thrust washers.

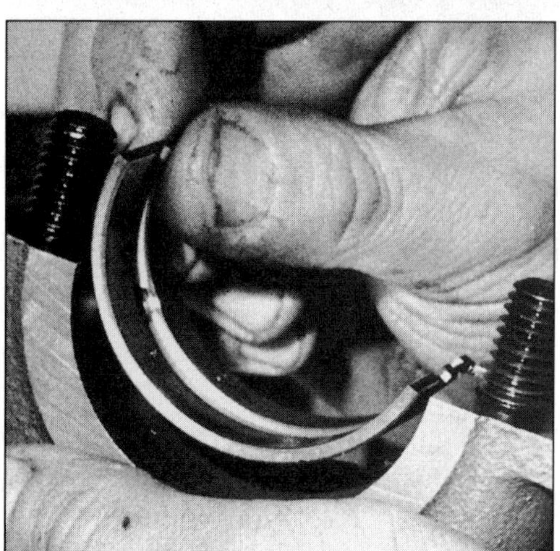

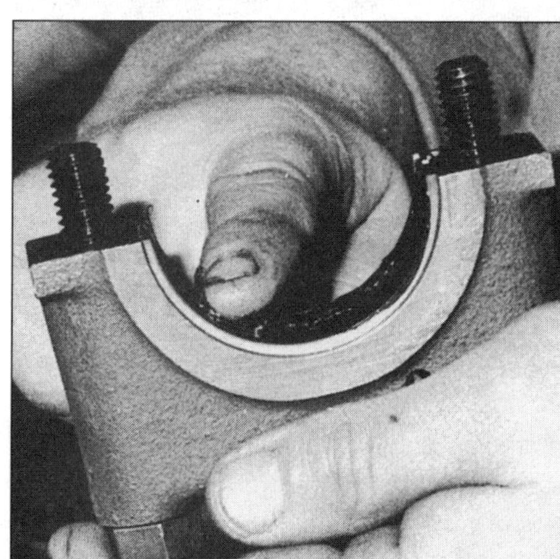

◀ Clean the main bearing caps and fit the shells.

◀◀ Put a smear of graphite compound on the shells.

▶▶ *Drop the bearing caps in place taking care to get them in the right order.*

▶ *Spin the nuts down with a speed brace and check that the crankshaft is free.*

▶ *Finally pull the cap bolts down to the specified torque.*

chances of a mistake when the shells were handed to you is greater than the chance of a mistake in the grinding. The oversize should be stamped on the outside of the bearing shell and will say +20, or +25, or whatever – the figures on an English or American shell indicating thousandths of an inch. On Continental made shells you may find the oversize in thousandths of a millimetre in which case you should have had the shaft ground to metric undersize, not Imperial.

Assuming the shells are the right size, smear a light layer of blue evenly over one of the big-end journals on the crankshaft, fit the appropriate con rod complete with its shell bearings and tighten the caps to the full torque of 31 to 35 lb ft (42 to 48 Nm). Now turn the connecting rod round several times on the journal and undo it carefully so as not to disturb the blue pattern. There should be a thin even layer of blue over the whole of the journal except, perhaps, the very ends by the fillet radius. Repeat this for each of the big-end journals and its rod.

If on any journal the blue is much darker one end than the other it means that the journal has been ground with a slight taper. Journal taper, remember, should not be greater than 0.0005 in (0.013mm) which would not show up as a darkening of the blue. If the blue has disappeared from any areas it means that there are high spots in the grinding. Clean the blue off and check again just to make sure but, if you find the same fault, take the shaft back to whoever reground it.

Do the same check on the main journals with the crankshaft mounted in the crankcase and the main bearing caps tightened to their correct torque of 55 to 60 lb ft (75 to 82 Nm). The only difference is that this time you check all the bearings at the same time. If, by any mischance, the shaft goes too tight to turn when you torque up all the bearing caps, one or more of the main journals may not have been ground to the correct undersize. Loosen and retorque each one in turn till you find the guilty journal.

One other possibility is that, with a Tuftrided shaft, the Tuftriding has distorted the shaft slightly and it has not been straightened properly. It should have been checked at the grinding stage but, if you get a tight shaft and can't find any other reason, have it set up between centres and checked with a dial test indicator.

It should go without saying, but I'll say it just the same, that the surfaces where the bearing shells fit in the connecting rods and in the crankcase and main bearing caps must be absolutely clean and that the locating tongues on the shells must fit properly in their slots in the rods, crankcase and bearing caps.

If all is well, clean every trace of blue off the crankshaft and bearing shells and put them away where they won't pick up dust and dirt.

The other way to check bearing clearances is with a very useful product called Plastigage. This is a thin round section strip of very accurately graded plastic which, when it's deformed, stays deformed and doesn't spring back. It's made by The Perfect Circle Corporation in Hagerstown, Indiana, USA in various sizes to suit different clearances. It's available in the UK in a handy kit, of the right sizes and sufficient quantity to rebuild an engine, from Burton Power Products Ltd, 631 Eastern Avenue, Ilford, Essex IG2 6PN.

To use it, say to check the bearing clearance of a big-end, you cut a length of Plastigage the same size as the length of the crankshaft journal bearing and lay it lengthways along the journal. Then you bolt on the connecting rod,

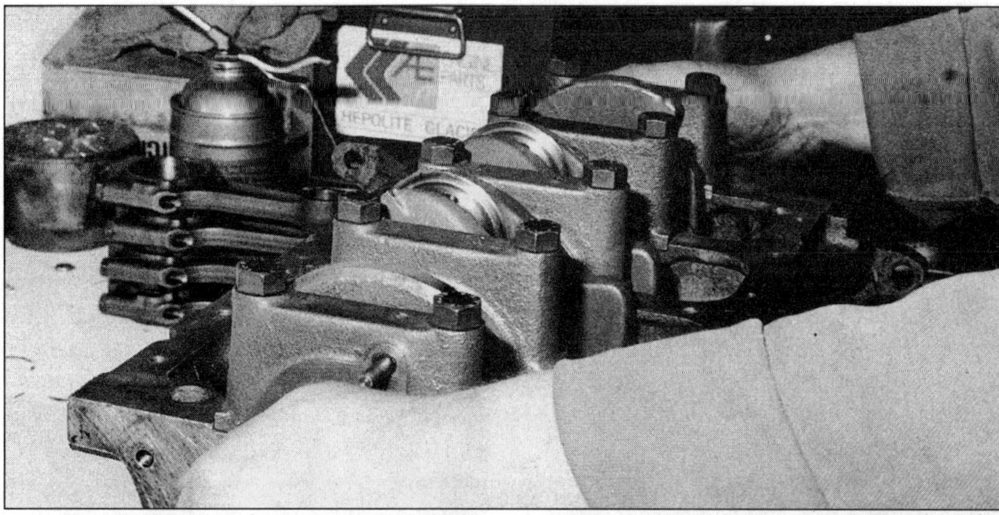

▶ *After torqueing down, check again that the shaft revolves freely.*

▶ *When you fit the rear cover, make sure the seal is concentric to the crankshaft. There is a Ford sleeve for this job, and you may be able to borrow one.*

▶▶ *Before you can move the gudgeon pins, the pistons have to be heated either on an electric heater (as here) or by putting them in very hot water.*

▶ *The pistons ready to receive the con rods.*

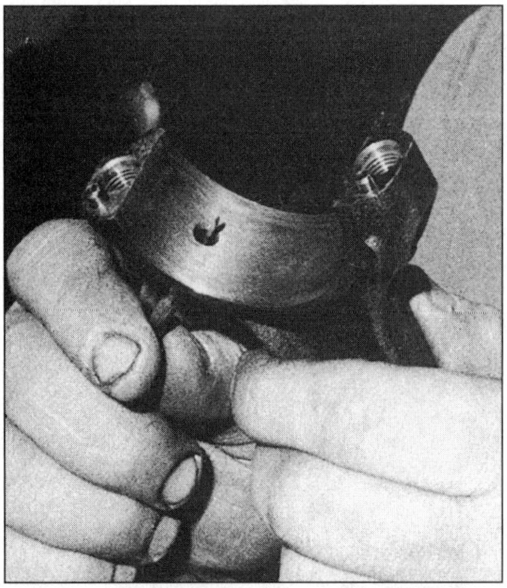

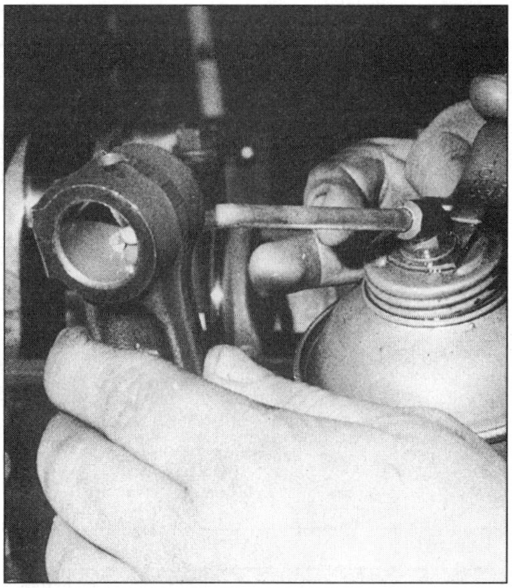

◀◀ *Check that the oilway in the big-end is free and clean.*

◀ *Squirt plenty of oil in the small-end bush …*

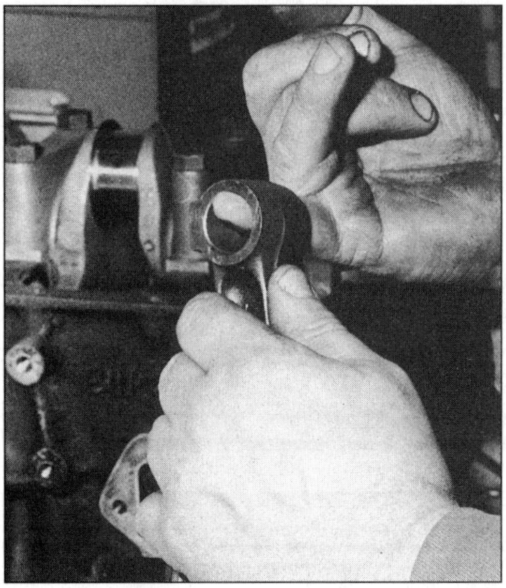

◀ *… or smear the inside with graphite compound.*

▼ *With the piston hot, align the con rod and push the gudgeon pin home with a shouldered mandrel.*

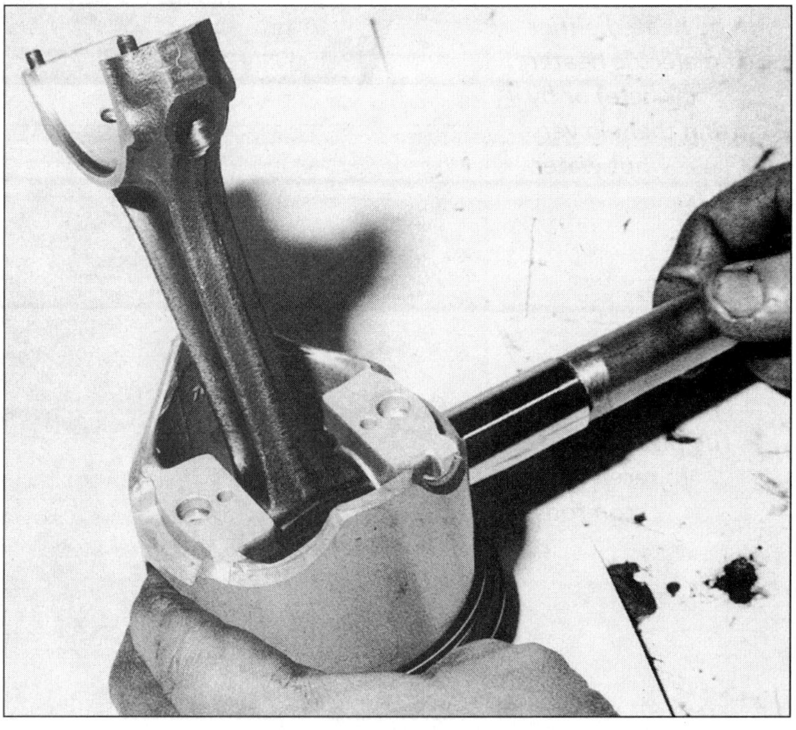

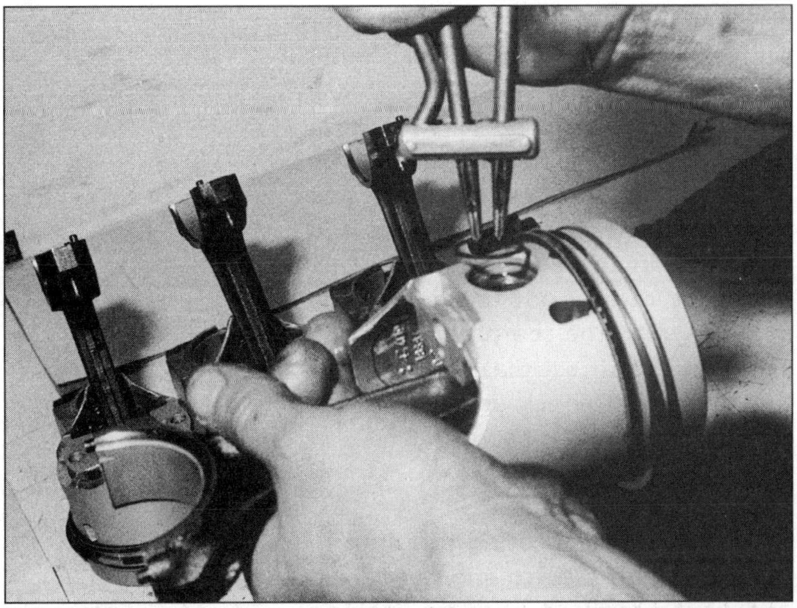

▲ *Make sure the end circlips are both secure.*

▶ *Squirt plenty of oil in the crankshaft oilways, then turn the engine over and support it on blocks to keep the crankshaft off the bench.*

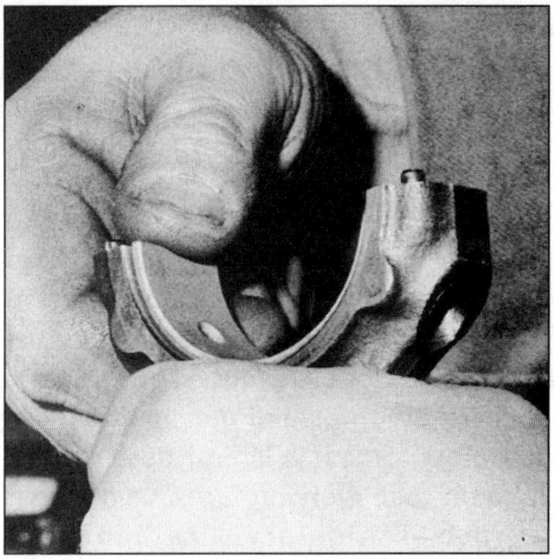

▶ *Smear graphite compound all round inside the bores before fitting the pistons.*

▶▶ *Clean the ends of the rods and fit the shells.*

complete with its bearing shells, and tighten the bolts to the full torque. Be careful not to rotate the bearing while you're doing this. When you undo the rod, the Plastigage will have squashed down to a flat strip.

Without taking the strip off the journal you compare the width of it with a series of measured widths on a gauge, rather like a large bar code, and the width on the gauge which matches the width of the plastic strip indicates the clearance.

You check the clearance of main bearings in the same way, checking one at a time, but with the other bearing shells in place to support the shaft. It's a very useful and clean way of checking clearances and, provided you tighten the cap down evenly and don't thump it down, it's a very accurate way. It will also tell you if the journal is tapered because, if it is, the Plastigage will be flattened wider at one end than the other.

When you're happy that all is well with the crankshaft, clean out the housing at the clutch end which takes the clutch shaft bearing and, if it's been renewed, make a final check by fitting the end of the crankshaft over the gearbox clutch shaft, input shaft or first motion shaft – whichever you like to call it. Have the gearbox firmly anchored down while you do this, or get someone to hold it, as crankshafts are heavy and, if you rock it, you could damage the needle bearing. Make sure you offer up the shaft straight and level. It's unlikely you'll find a problem but, if you do, it's better to discover it now rather than later when you're trying to fit the gearbox on.

Now you're ready to begin the actual assembly. Coat the sides of the cam followers with oil and put them back in the block. Smear a liberal dose of cam lube on their working faces. Cam lube is a special high-pressure lubricant containing zinc dio-thio-phosphate for running-in camshafts, and you should have got a bottle of it with your new camshaft kit. Accept no substitutes, as the adverts say. Then smear cam lube on all the working surfaces of the cams and liberally oil the camshaft bearings and bearing journals before you slide it in. Again, be careful not to damage the bearings with the sharp edges of the cams. Fit the thrust plate with a new locking plate; check again that the camshaft end float is between 0.0025 in and 0.0075 in (0.064mm to 0.191mm) and lock the bolt heads.

Clean the main bearing housings in the block with a petrol-moistened rag and fit the shells in position, making sure that you get the locating tags positioned correctly. Give the surfaces of the shells another wipe, just to make sure they're clean, then liberally oil the faces of them (or smear them with a graphite running-in compound) and lower the crankshaft into position. Fit the thrust half-washers which go either side of the centre main bearing with their oil grooves facing the crankshaft flange. Check the crankshaft end float, which should be between 0.003 in and 0.011 in (0.075mm to 0.280mm). If it's outside these limits you should be able to get different thickness thrust washers, either from a Ford dealer or from an engine specialist. There are only two thrust half-washers and they fit in the crankcase. Thrust washers are not fitted each side of the main bearing cap.

Coat the faces of the bearings quite liberally with oil or running-in compound and make certain you fit the end caps in the right order and the right way round. Check again that the threads of the cap bolts are clean, give them a light coating of oil and torque them down to 55 to 60 lb ft (75 to 82 Nm). Give the crankshaft a few turns to check yet again that everything is free.

▶▶ *Once again, use graphite compound on the shells.*

▶ *Check you have the pistons in the right order, and lower the rods down into the bores.*

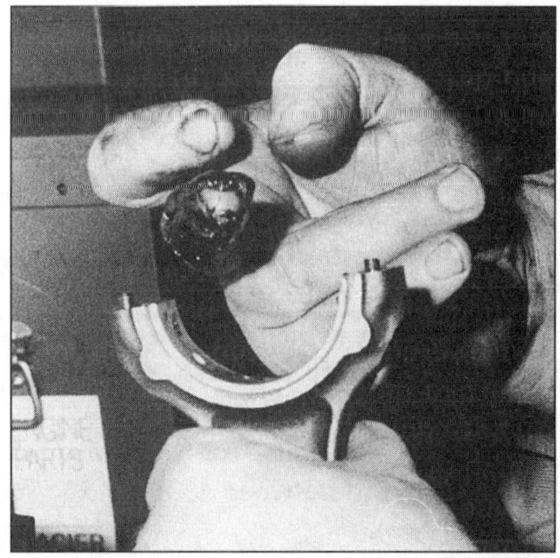

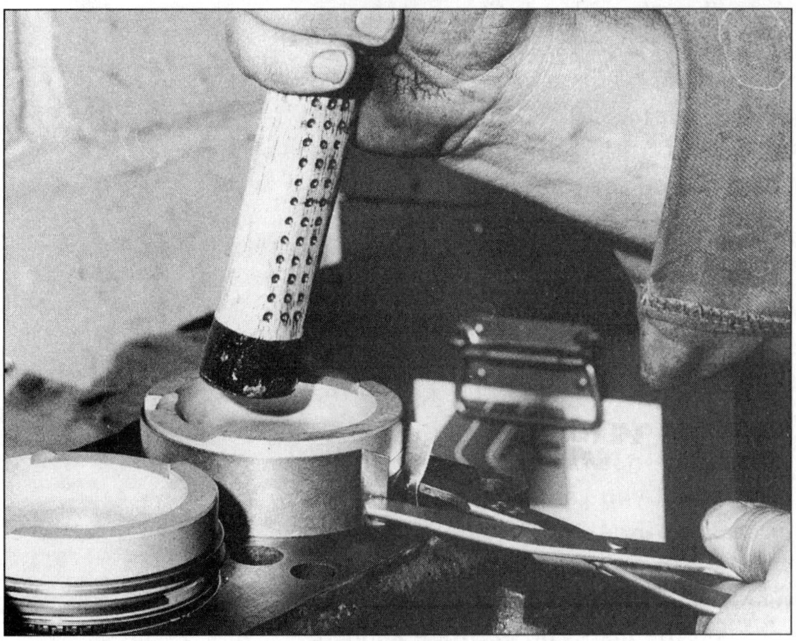

▲ *Use a piston ring clamp and tap the pistons down into the bores.*

▶ *Turn the engine over again, fit the big-end caps, spin the bolts down and torque them.*

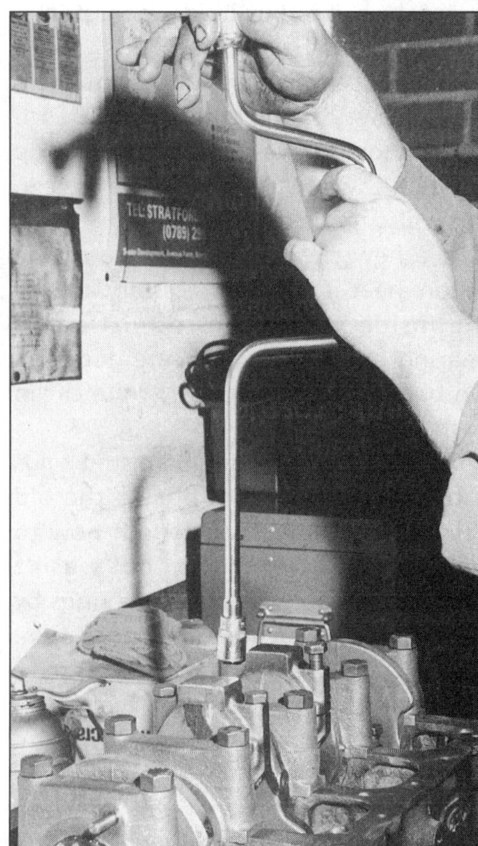

Fit a new oil seal into the rear cover and lubricate the lip of the seal before you slide the cover on. To avoid oil leaks into the bellhousing it's important to fit the cover with the oil seal centralized. It's possible, with luck, to centralize it by eye, but much better – if you can get hold of one – is to use the Ford tool. This is a sleeve which fits over the end of the crankshaft and into a recess in the rear cover to hold it central while you tighten the bolts. The bolts should be tightened to between 13 and 15 lb ft (17 to 21 Nm).

Next we come to fitting the pistons on their connecting rods. The gudgeon pins are a tight press fit in the pistons which have to be heated before the gudgeon pins are moved. You can heat them in hot water – or, if you have one, on an electric hotplate – until the pins press out easily. There's no need to take the pin right out, just push it out far enough to get the con rod up inside the piston.

Before you fit the con rods, make absolutely certain that you've got them the right way round in the pistons. Each piston should have an arrow on its top pointing towards the front, and the con rod is also marked 'Front'.

Fit a new circlip into one side of the piston, heat it again if necessary and oil the inside of the con rod little-end bearing before you push the gudgeon pin back. When it's home, fit the circlip to the other side.

The easiest way to fit the piston rings is to use a proper tool but, in the old days, an apprentice was taught how to fit them by hand if the tool wasn't available. If you want to fit the rings by hand, practice on some old ones first. Once you're used to it, the job's quite simple but, until you get the feel for it, you can break rings very easily. The technique is to hold the ring with your thumbs on the edge of the gap and your two middle fingers supporting the other side. Just a light pressure with your thumbs is enough to open the ring sufficiently to fit it over the top of the piston and down into its groove.

When you fit the rings make sure that the compression rings are the right way up. Plain rings will have the word 'Top' on the top face. If you are fitting new rings to a worn bore and using a ridge-dodger top ring, you can identify the top by the small ledge. Space the ring gaps at approximately 120° round the piston.

Feed the pistons and con rods in from the top of the bore, checking again that they are the right way round, coat the rings liberally with oil, fit a piston ring compressor and tap the piston down into the bore. I generally use the handle of a hammer for this. Turn the crankshaft till the journal to which you're fitting the rod is at the bottom of its throw, fit the shell bearing in the rod, coat it with oil and pull it into position. Fit the other half of the shell bearing in the cap, fit it on and torque the bolts up to 31 to 35 lb ft (42 to 48 Nm). Fit a new Woodruff key to the nose of the crankshaft.

If you are not fitting a vernier timing chainwheel you can time the camshaft before going any further. Fit the crankshaft and camshaft chainwheels temporarily without the chain and turn them till the marks on each are aligned. Then take them off again and refit them together with the chain. Remember to lock the bolts of the camshaft chainwheel. Fit the chain tensioner then compress the spring while you slide the tensioner arm on to the pin at the front of the main bearing cap. Then just release the spring. There's no adjustment to be made.

Press a new seal into the front timing cover and smear oil on the inner lip before you slide the cover over the nose of the crankshaft. On earlier engines you have to take the same care as with the rear cover to centralize the oil seal

▲ *With the pistons in place, the engine will be getting stiffer to turn and you may have to use a bar to check that everything is free.*

▶ *If you are not using a vernier chainwheel, you can fit the timing gear with the engine upside down*

▶ *Check that the timing marks line up before you tighten the bolts.*

▶▶ *Turn the crankshaft a couple of times, recheck the timing and then lock the tab washers.*

◀◀ Screw the new chain tensioner in place.

◀ Make sure the tensioner spring is hooked in place and let the cam go to take up the chain slack.

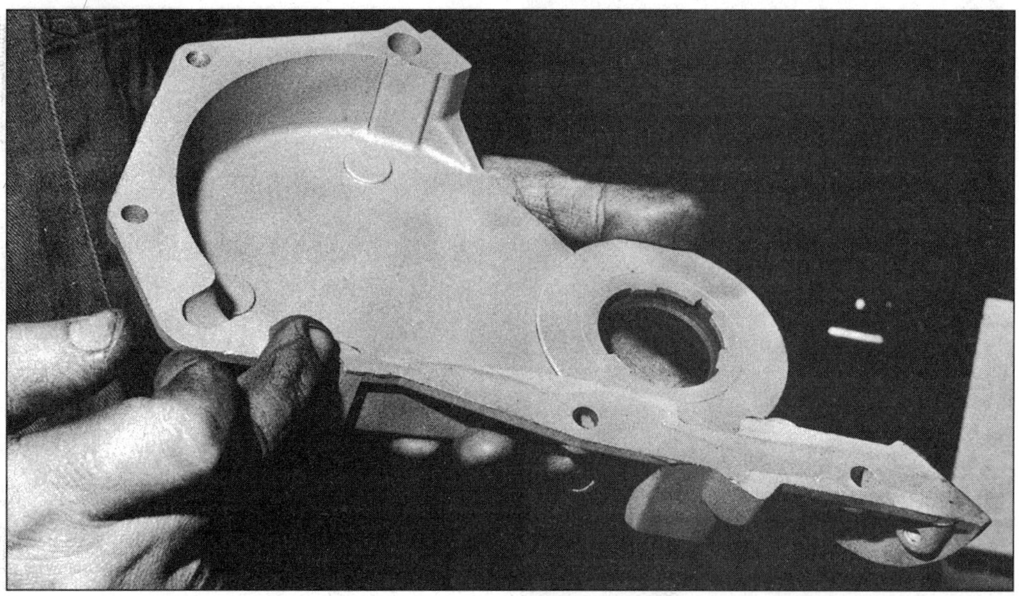

◀ There's no need to use compound on joint faces, just a smear of heavy grease will give an oiltight joint.

▼ Be careful not to damage the oil seal when you fit the front cover.

▲ *The two bolts on the front cover nearest the sump flange are dowel bolts. Make sure you use the correct ones, and tighten them first to position the cover.*

▶ *Smear the tube of the oil pick-up pipe with Loctite before fitting it.*

▲ *Check that the pick-up pipe is not hit by the crankshaft when it revolves, and that it is free of the sump.*

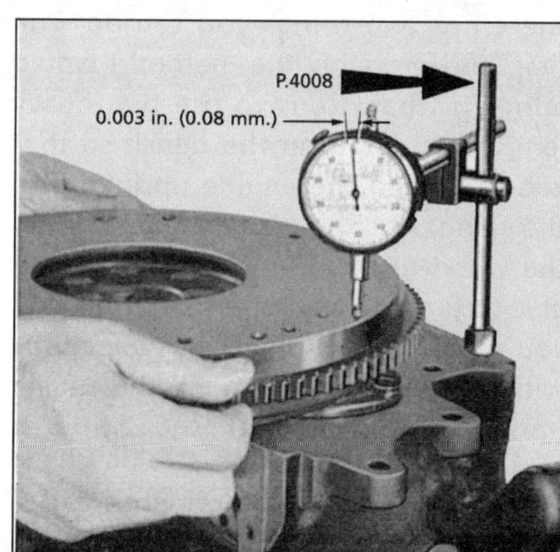

▶ *If you have a dial test indicator (DTI), check the flywheel run-out. It should be within 0.003 in total indicator reading.*

but, on later engines, the two bolts nearest the sump flange are dowel bolts and align the cover. Fit these first. Remember, before you fit the timing cover, to put the oil slinger on the nose of the crankshaft with its concave side facing outwards.

Some people are never happy building an engine unless they smear every gasket with jointing compound. If you checked and dressed all the mating surfaces as I suggested earlier there's no need. I use jointing compound only in one place, on the rocker cover, as I'll explain below. All you need to provide a completely oil-tight joint is a smear of heavy grease on the gaskets. I use a heavy yellow waterproof grease made for water pumps.

If you're fitting a vernier timing chainwheel, you adjust this with the inlet valve of No. 1 cylinder on full lift, as I described in Chapter 10. Unfortunately, on the Crossflow, the front of the sump fits on the front timing cover which needs to be off to get at the chainwheels. Standing the engine the right way up, to check the push rod lift, is awkward as there isn't a flat base on which to stand the engine. This is where an engine stand is invaluable – you just turn the engine over at will.

If you haven't got an engine stand there are two things you can do. The first, and probably the method I would adopt if I had to, is to use two heavy lengths of wood on the bench so that you can stand the engine upright with the flanges of the crankcase resting on the wood to lift it up sufficiently to be able to turn the crankshaft. Alternatively, you could bolt the sump on temporarily without its gasket and stand the engine upright on this. Neither is ideal, but it means the job can be done.

You may have seen a racing Crossflow where someone has modified the front timing cover by cutting a hole to get at the camshaft chainwheel, drilling and tapping six holes round this and making a plate and gasket to close it afterwards. This wasn't done because the engine builder didn't have an engine stand, it was done so that he could adjust the cam timing to get the best power curve with the engine installed and the car on a rolling road. I wouldn't recommend doing this in a home garage, as some of the ribbing on the front of the timing cover has to be machined off to get a flat face. You might, though, if you want to do this, be able to have a cover modified by one of the engine tuning firms or a machine shop.

Some vernier chainwheels have basic timing marks scribed or indicated by dots on the teeth. The adjustable centre plate isn't usually marked, but you set it so that it is in the middle of its movement either way in the slots. This lets you set the timing near enough to what it would be with a standard non-adjustable chainwheel and gives you a basis from which to work. Some adjustable chainwheels aren't marked, so if yours isn't marked, set the basic timing with the original Ford chainwheels (no need to put the chain on) then centre the adjustment of the vernier chainwheel in its slots, feed on the two chainwheels complete with the chain and lock the camshaft bolts.

Turn the engine upside down again, fit the oil pick-up pipe after smearing the tube lightly with Loctite and bolt its bracket down. Turn the crankshaft to make certain that the pipe is clear of the webs, and drop the sump on without its gasket to make sure that the pipe hasn't been bent so that the strainer fouls the bottom of the sump.

It makes torqueing up the flywheel bolts and the nose pulley bolt easier if you put them on before fitting the

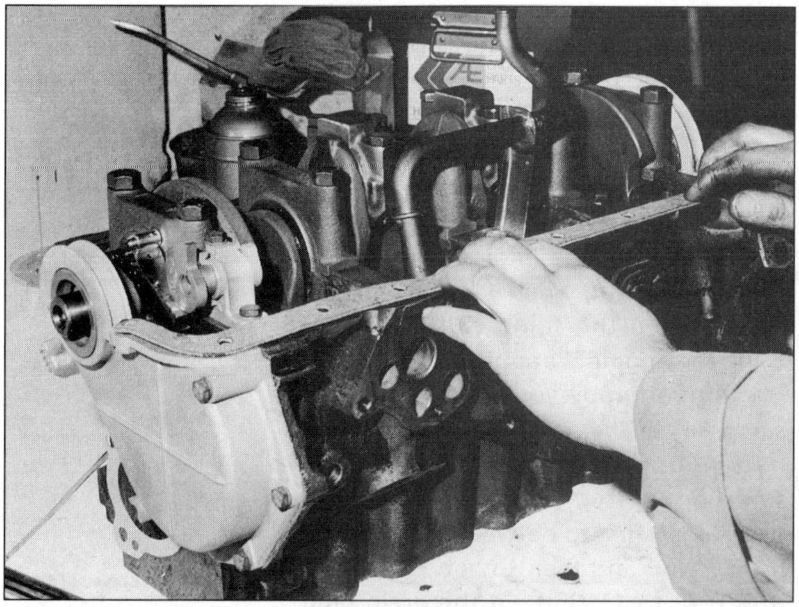

▲ *Put a smear of grease on the crankcase flange, and position the sump gaskets on each side.*

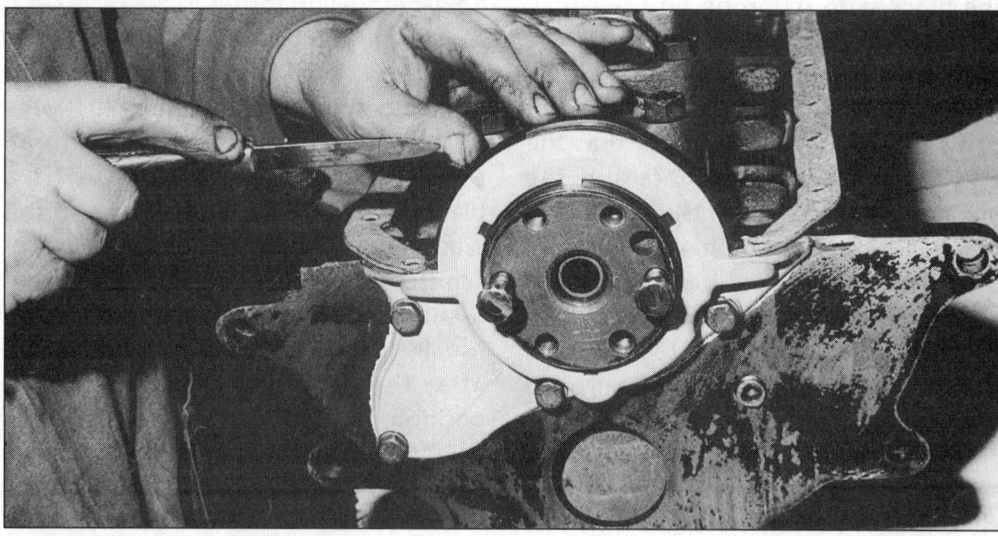

▶ *The end seals fit in the cover grooves. If necessary, ease them in with a knife blade.*

▶▶ *Tighten the sump bolts diagonally so that it seats evenly.*

▶ *With the sump on, you can turn the engine over again. You may like to use a block of wood under the front to level it.*

sump, as you can use a block of wood between the rear crankshaft web and the side of the crankcase to stop the shaft rotating. Torque the nose pulley to between 24 and 28 lb ft (33 to 38 Nm) and torque the flywheel bolts in a diagonal pattern to between 50 and 56 lb ft (68 to 76 Nm). Make sure that you put the flywheel on in its balanced position, lining up the marks made by the firm that balanced the flywheel and crankshaft.

It's not very likely that the flywheel run-out, a buckled movement of the clutch face, will be outside limits but, once again, it does no harm to check. If it is out, it can lead to early clutch problems. Use a DTI on the rear face of the block with its probe on the clutch face of the flywheel, and turn the crankshaft. It should not be greater than 0.003 in (0.08mm) total indicator reading. If it's outside this limit, check that the end face of the crankshaft and the mating recess in the flywheel are clean. If that doesn't cure the problem, check with the person who refaced the flywheel and ask him to double check his work. I have come across only one refaced flywheel that was bad, but that was a really bad one, nearly 0.018 in run-out.

As you had the clutch balanced separately from the crankshaft and flywheel, you can bolt it on in any position using either a proprietary tool or an old gearbox clutch shaft to align the splines in the centre plate. Tighten the bolts diagonally to pull the springs down evenly, and torque them to between 13 and 15 lb ft (17 to 21 Nm).

If you happen to be building an engine up with components from several, make sure you use the correct sump. A number of sumps were used on the Crossflow, some with the deep pan at the front and some with it at the rear. You also have to be careful that, on a 1600 engine you don't use the sump from a smaller one. The 1300 sump, for example, fits on the 1600 crankcase and all appears to be well. There isn't, however, sufficient clearance for the bottoms of the con rods with the longer throw 1600 crankshaft. The interference is so slight that you may not feel it when you turn the engine over during the building but, when you start up, the clatter from the sump will make you think the engine is about to fly apart.

If you're happy you've got the right sump, smear the gaskets and the sealing half-rings with grease and bolt the sump down, tightening diagonally in two stages, first to between 3 and 5 lb ft (4 to 7 Nm) and then to between 6 and 8 lb ft (8 to 11 Nm). Fit the drain plug now, before you forget it.

Now you can turn the engine the right way up and fit the oil pump (remembering to fill it with oil first), the petrol pump (unless you're using an electric pump, in which case fit a blanking plate), the dipstick tube (if this was taken out) and the oil separator canister (unless you're using a different crankcase breathing arrangement). I usually leave fitting the distributor until after the head and rocker shaft is in position, for two reasons. First, it's quite vulnerable sticking out of the side and, second, it's very easy to time it on the wrong stroke with the crankshaft 180° out if you can't watch the valves. In the meantime, I stuff a large piece of rag in the hole to stop anything falling in. I once had to partially strip an engine again because I dropped a small spring washer and, with the whole garage floor in front of it, it had to drop straight down the open distributor hole.

If the valves and seats were refaced by a tuning firm with sophisticated equipment, the seating will have been tested with a vacuum pull-down and there should be no need to lap them in but, with firms that I don't know well, I

▶ *Prime the oil pump by filling it with oil before you bolt it in place.*

▶▶ *Use dowel bolts to position the head gasket.*

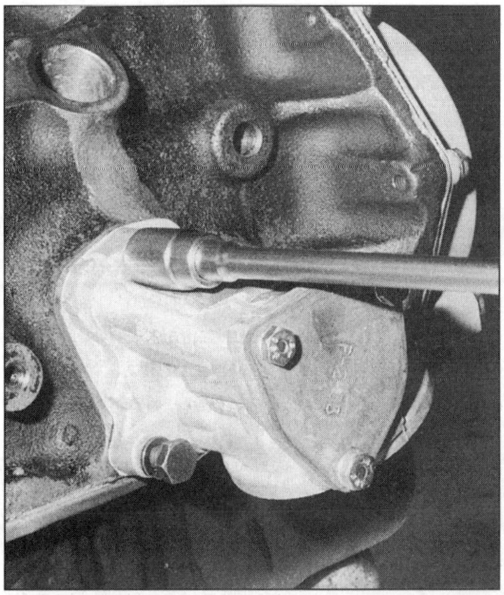

▶ *Torque down the head bolts in the correct order and sequence.*

▶ *The correct head tightening order.*

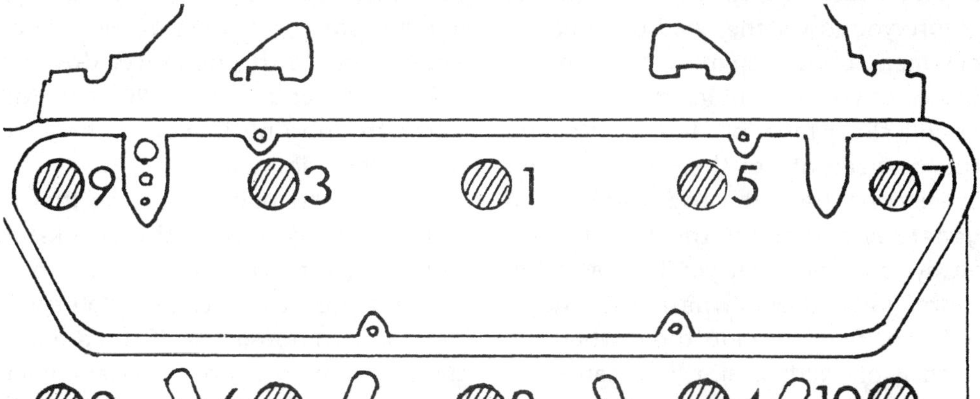

never feel quite safe without a light lap. If they were done by a general engineering firm I would definitely advise light lapping in with fine grinding paste. As you lap each one in keep it in order in a strip of wood with eight holes in it. Make sure every trace of the paste is cleaned off afterwards. A fine line of lapping is sufficient for the inlet valves but, to help them get rid of their heat, the seating on the exhaust valves should be reasonably wide.

When you fit the valves there won't be room for the standard Ford shroud type oil seal if you are using double valve springs. Some people put small O-rings on the valve stems in place of the shrouds. I haven't got a great deal of faith in them but, however, I have found that the shroud type seals from a Metro will generally clear the inner spring. The inlet valves are the important ones to be fitted with oil seals to avoid oil being sucked down into the combustion chamber on the inlet stroke. When the valves are all in position I generally go along them and bounce a fairly heavy hammer on the top of each, just to make sure that the valve collets have seated properly. It isn't strictly necessary, but it gives peace of mind. Remember to oil the valve stems liberally before fitting.

Because the Crossflow uses head bolts and not studs, you need a couple of dowel bolts to position the gasket before you lower the head on. You can buy them, or you can make them from a couple of old head bolts. If you do this, grind a slight taper on the top after you cut the head off, so the gasket doesn't catch on the cut, and hacksaw a generous slot across the top. There's always a chance that you'll screw them in too far and won't be able to get a hold on them to undo them when the head is on. With a slot, you can undo them with a screwdriver.

Before you fit the head, make sure that the bolt holes in the block face are clean and dry. If they're dirty, it will upset the torque reading and, if there's oil in the bottom, you could get a hydraulic lock and reach the torque figure without the bolt clamping the head down properly. We usually run a tap down them as a check on the threads and then blow them out with compressed air. If you haven't got a tap, you can use an old head bolt with a flat ground or filed on two sides of the thread to make sure they're clean.

Depending on the make of gasket you are using, the recommendation might be to fit it dry or with a smear of grease. Always follow the gasket maker's instructions. Before you fit the cylinder head, turn the engine until all the pistons are about halfway down the bores, and run a ring of oil round the top of each piston to help the initial bedding. Turn the engine over a few times to distribute it.

Put the bolts in finger tight, noting that two of them, Nos. 4 and 6 in the tightening diagram, are shorter than the others. Follow the tightening torques carefully, always using the tightening sequence in the diagram. The first tightening, to seat the bolts, is to 5 lb ft (7 Nm). Then torque down to between 20 and 31 lb ft (28 to 42 Nm), then to between 52 and 56 lb ft (70 to 76 Nm). Wait between 10 and 20 minutes for the stretch in the bolts to even, then pull the bolts down to between 66 and 71 lb ft (90 to 97 Nm). After the engine is warm, check the final torque figure again.

Fit the water pump, checking that the pulley is in line with the crankshaft pulley. When you fit the thermostat housing you have a choice. You can fit the standard housing or, if you are using the engine with a cooling system header tank, you can fit the thermostat housing used when the Crossflow is fitted to pre-

▶ *Fit the push rods in place before lowering the rocker assembly down.*

▼ *Tighten the rocker pedestal bolts gradually and evenly, as some valve springs will be in compression.*

▶ *Set the valve clearances before setting the ignition timing or, if you use a vernier camshaft chainwheel, before finally timing the camshaft.*

▶ *It helps to spread the load of the bolt and avoid leaks if you use the banana-shaped rocker cover bolt washers from a Sierra.*

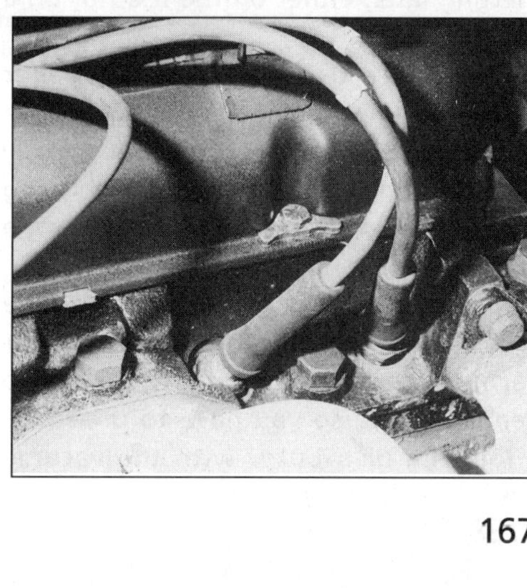

1983 Fiestas. This has a pressure cap on the top for filling the system. There is another alternative. If you want to fit an electric cooling fan to the car you can use the thermostat housing from the Crossflow fitted to the Mk I XR2 Fiesta. This has a mounting on the top to take the fan's thermostatic switch.

Put the push rods in position, assemble the rocker gear on its shaft with plenty of oil and, when you fit it, tighten the bolts down evenly and gradually because some of the valve springs will be in compression. Finish off to a torque between 18 and 22 lb ft (24 to 30 Nm). Set the valve clearances and turn the engine over several times to make sure everything is free.

Now you can fit the distributor and time it. Turn the engine till both valves of No. 1 cylinder are closed, so you can move the rockers and the piston is on TDC. Then, because the static timing fires the ignition before TDC, turn the crankshaft back to the appropriate point. If you're using the standard front pulley it will have timing marks on it but, if you're using a plain steel one it won't, so you'll have to use your timing disc. The best static ignition timing may vary depending on the type of performance camshaft you have used and even on the gas flow porting of the head, but a good starting figure is 10° before TDC. You can check the final setting later, either on the road or on a rolling road.

If you are using a contact breaker distributor, set the points gap to 0.025 in (0.64mm). You can set it more accurately later, when the engine's running, using a dwell meter to give a dwell angle between 48° and 52°.

Because the distributor is driven by a skew gear on the camshaft, the rotor arm turns as you feed the gear into engagement, so you have to pre-set it. Hold the distributor with the vacuum unit pointing towards the rear of the engine and at about 30° to the block. Turn the rotor arm so that it points towards No. 2 inlet port and lower the distributor down into engagement with the camshaft. As the gears engage, the rotor arm will turn until it is almost at right angles to the block. Fit the bolt which holds the distributor clamp to the block and tighten it, but don't tighten the bolt which clamps the distributor body.

Turn the body of the distributor so that the rotor arm is pointing towards No. 1 stud in the cap and, with a contact breaker, so that the points are just about to open. If you are using the Bosch electronic distributor the setting will be so that the trigger wheel and stator are in line. This will give you a basic ignition timing which can be adjusted later when the engine is running.

As I mentioned above, the rocker cover is the only place where I use jointing compound – not to prevent leaks but to stick the gasket to the rocker cover. Then, when you have to lift it to adjust the tappets at any time, the gasket stays on the cover and you can use it again without trying to position it on an oily surface. Quite a few oil leaks at Crossflow rocker covers are caused by overtightening the screws which hold it down. Part of the blame, however, must go to Ford for using just a small washer under rocker cover screws. The localized pressure raises bumps inside the rocker cover and it leaks between the bumps. They realized their mistake much later and introduced elongated washers on the Sierra. Get yourself four of these and use them on your Crossflow to spread the bolt load.

Whether you fit the inlet and exhaust manifolds, starter motor, alternator and the other bits and pieces now or when the engine is in the car is a matter of choice.

Chapter 16

Overhauling a Weber DGAV 32/36

Webers are, without doubt, the most popular carburettors with people who tune Crossflows. They give better performance and economy than the Motorcraft, and though they might look complicated they're so much easier to overhaul and set up than Ford's notoriously tricky VV (Variable Venturi). You can get overhaul kits for almost all Weber carburettors from any Weber agent.

▲ *You can pick up Weber DGAV carburettors quite cheaply in breakers' yards.*

Tuners who are going for racing powers often use a pair of massive Weber DCOE 40s or 45s but, for the majority of people who tune their Crossflows for fast road work, often doubling at times as a family car, the DGAV 32/36 or DGV 32/36 is a favourite. They are very similar, the main difference being that the DGAV has an automatic choke, whereas the DGV uses a manual choke.

A lot of people don't like automatic chokes. They will tell you that they are unreliable and that they worsen your fuel consumption. I must say that I haven't found this to be the case. When a Weber automatic choke is properly set up it works well.

If you are on the look-out for a second-hand carburettor to overhaul, you will probably be offered a DGAV. If you finally decide that you don't like an automatic choke you can quite easily fit a manual choke conversion.

The Weber DGAV 32/36 was fitted to some sporting versions of the 1600 Escort and a large number of 1600 and 2-litre Cortinas and Capris and, though these will be jetted and calibrated for an overhead camshaft Pinto engine, any Weber agent will be able to supply you with suitable jets for a 1300 or 1600 Crossflow, standard or tuned. With a tuned engine the jets will be OK to get it started and the car running, but your first stop should be a tuning firm with a rolling road. They will be able to plot the mixture strength over the whole of the rev range and fit any different-sized jets that may be necessary.

Don't go for full power readings on a newly rebuilt engine. Wait until it has a few thousand miles bedding-down under its belt, then take it back to the rolling road for a final check.

You will also need a manifold to go with the Weber because the Crossflow manifold which takes a Motorcraft or VV will not accept a Weber DGAV 32/36. You can get a suitable manifold from Weber or from a number of specialist tuning firms, or you can look out for one from a Mk I or Mk II Escort GT, or a Crossflow-engined Mk I Capri GT or a Mk II Crossflow-engined 1600 or 2000 Cortina GT.

These all used twin-choke Webers, and the manifolds will fit physically on to any capacity Crossflow. What you will have to watch is matching the ports in the head to those in the manifold, which was covered in Chapter 4.

If you are hunting for a carburettor in a breaker's yard, don't necessarily grab the first Weber you find, or you might end up with one that's so worn or corroded inside that it's useless. Even if it means hunting around for a week or two, make sure you get a good one. There won't be any difference in price at the breakers.

How do you tell whether it's good or not? One cause for rejection is a throttle spindle that waggles about in the carb body. The spindle itself doesn't usually wear so much the wear is in the body of the carb, and there's precious little you can do at home to remedy it.

Don't try drilling and reaming it to fit bushes unless you've got a very well-equipped machine shop. The chances of getting the bushes in line with a hand reamer are almost nil, and you'll never get the throttle to close properly.

You may find your carburettor in the type of breaker's yard where you are expected to walk round and take components off yourself. In which case, before you take the carburettor off the car, release it from the pedal mechanism and try wobbling the throttle spindle. The primary throttle, the one the pedal's connected to, invariably wears more than the secondary. Any carburettor with thousands of miles under its throttles will have a certain amount of wear at the spindle. What you have to decide is whether the wear is too much.

I can't tell you how much is too much. A carburettor specialist will say that any wear is too much, and I would certainly reject any carb where the spindle moved 1/32 in or more in total. But I have successfully overhauled carbs with just a few thousandths of an inch rock at the spindle, and they've gone on to cover thousands of miles with no slow-running or other mixture problems. It's a case of try it and see, and use your own judgement.

In any case, wear at the throttle spindle doesn't affect the mixture very much except at tickover with the throttles closed. The wider you open the throttles the less it affects the mixture, partly because the volume of air passing through the carburettor is large compared with any leak, but mainly because, once the throttle starts to open, the depression at the throttle spindles drops rapidly, so less air is sucked in round the worn spindle.

Another case for rejection is the carburettor that's been standing for ages with the top open to the elements. This generally makes them difficult to overhaul, sometimes to the point where it isn't worth it. The throttles will be seized and stiff and there will be so much dried muck in the internal drillings it will take a month of soaking and blowing through with an airline to get it clean.

Webers have lead plugs in them which can be taken out to get a cleaning rod through all the drillings. But, though Weber will supply new ones, these plugs are not openly listed in the catalogue and Weber does not recommend that they are fitted without the special staking tool. If they are not fitted properly you get fuel leaks or air leaks that can upset the performance, and which can be very difficult faults to track down. Try if you can to get a carburettor from a car which has been running recently.

The carburettor which was photographed for this chapter came from a two-litre Cortina that had been running only a fortnight before. There was a small amount of wear at the throttle spindle, but this was acceptable for the purpose (not, though, for top-notch racing tune), and the air cleaner was still in place, so it wasn't full of rain-water.

The first job with any second-hand carburettor is to clean all the oil and muck off the outside. There are several liquids you can use to clean it. Ordinary white spirit from a DIY shop is cheap, and good. You just dunk the whole carburettor in a bath of it, leave it for a day or so and then scrub away with an old toothbrush.

You could use a couple of aerosols of proprietary carb cleaner, but it's an unnecessary expense, except for the internals where it is invaluable. For this carburettor, for speed, I used a cheap form of cellulose thinners sold as gun wash. It whips the old muck off like magic but, if you do use it, work outside in the open air, and wear a filter mask (available from where you buy the gun wash).

After cleaning, the first job is to identify the carburettor accurately so you can get the right overhaul kit. It's no good just saying 'It's a DGAV'. There have been lots of models of DGAV over the years, so you need the other identifying numbers. You'll find them all stamped on the carburettor main flange. Mine was a DGAV/3D1. It does no harm to quote the serial number as well but the Weber agent probably won't need this.

Carefully save all the washers, circlips and other bits and pieces you take off. You'll get new circlips, split pins and rubber O-rings in the overhaul kit, but little circlips have a habit of pinging off across the garage, and it's handy to have a few spares. It is useful to replace all the screws in their holes after you take something off. That way you don't lose them or get them muddled up.

The first job is to check that the base flange is true. If it's warped you'll get air leaks. To true it after dismantling, rub it on emery paper held against something flat, like a piece of plate glass.

Start dismantling by prising off the small circlip and lifting off the top choke operating link – the one that's attached to the throttle spindle. Then you can undo the six screws holding the top of the carburettor and lift it off. Put the main body to one side for the moment.

Inside the top you'll find the float with its needle valve. You take the float off by pushing out its hinge pin, and then the needle valve just lifts out. If you've got an overhaul kit don't bother to inspect the needle valve or the brass valve seating into which it fits. You get new ones in the kit, but you don't get a float hinge pin so save that carefully.

Just in front of the needle valve is the power valve, held by three screws and with a tall spring sticking out the top. Take this off carefully because you don't get a new one in the kit. If the diaphragm is OK and the valve spindle isn't seized, you can use it again. If the diaphragm's damaged you can get a new valve from your Weber agent. You'll find the part number on the leaflet that goes with the kit.

Behind the float, under a hexagon nut, is the fuel filter. This used not to be included in the overhaul kit, but it is included in the later ones. Check the kit when you buy it.

▲ *Check the throttle spindles for side play. If there is too much wear, reject the carburettor.*

◀ *When you go for an overhaul kit don't just say: 'It's a Weber DGAV'. There were lots of different models. Quote the full number which you'll find stamped on a pad on the base flange.*

◀ *The first job is to check the base with a straight-edge to make sure it isn't warped. If it is, true it up with emery paper on a sheet of plate glass.*

◀ *Start dismantling by undoing the small circlip and lifting off the choke operating link.*

▶ *Then you can undo the six screws which hold the top of the carburettor to the main body.*

▼ *After the screws are taken out, the top lifts off complete with the twin floats.*

▶▶ *The pin which holds the floats might be tight, but it's only a push fit.*

▶ *In front of the float is the power valve, the one with a long spring on top.*

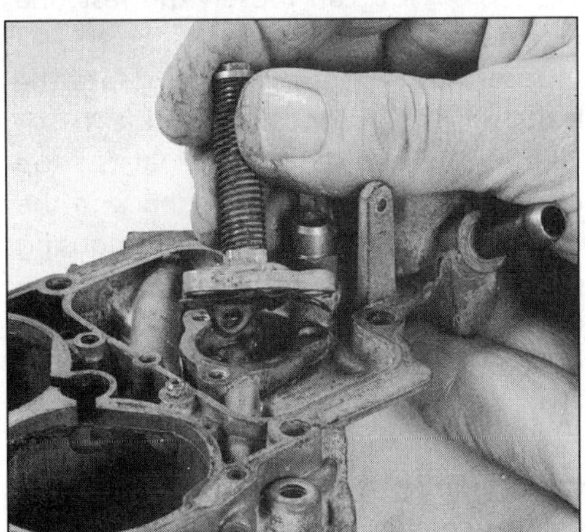

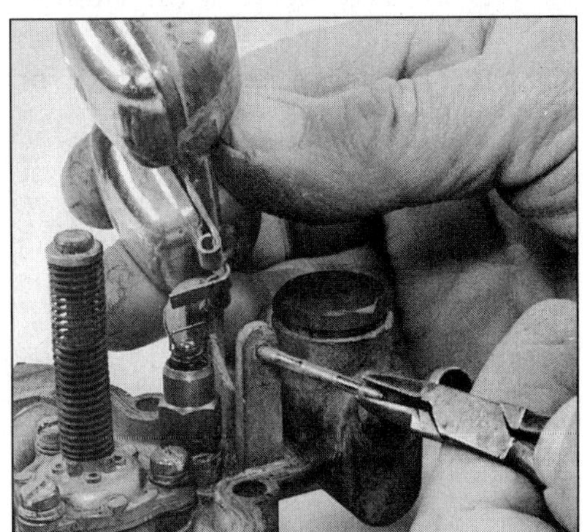

Once you've got all the components off the carburettor top you can soak it in cleaner once again to loosen the muck, and give it a good scrub off. If you've got an air-line, blow through all the holes and passages to make sure they're free. If you haven't got an air-line, you can get a useful jet of air by clamping the body of an old ball point pen in your footpump line. Follow up by squirting through all the passages with an aerosol of proprietary carburettor cleaner. Once again, be careful about breathing the fumes. They aren't very nice.

Now turn your attention to the main body and take out all the jets. Start with the two on the top of the body – the primary and secondary air correction jets. They've got brass emulsion tubes under them, and these sometimes stick. The easiest way to get them out is to ease them up with a small piece of wood. Don't use a screwdriver or you might damage them.

Keep all your jets strictly in order so you get them back in the same places. An easy way of doing this is to push them in a soft piece of expanded polystyrene, like an old ceiling tile or the packaging from a box. Later, when you've cleaned them, you can identify them by their numbers, but it's sometimes difficult to read these when the jets are dirty. Keep all the washers with the jets, even if you get new ones in the kit. They're useful for identification.

Next to the air correction jets is the accelerator pump jet which holds down a Y-shaped delivery piece overhanging the main venturis. This has got tiny little holes in it, and they must be free.

Now look down in the float bowl. The two main jets – primary and secondary – are set at an angle in the back of the chamber. The big one on the floor of the chamber is the full power needle valve, and its needle must be free inside the valve.

While you're at the top of the body there's another piece to take out and clean. It isn't a jet, it's the brass blanking plug for the accelerator pump and it sits in the top flange of the float chamber.

Next, go down to the side of the body where you'll find two fairly large brass screws, one each side. These are the primary and secondary idle jets and, when you take them out, you'll find they're in two pieces – a jet and a jet holder.

The last brass piece to take out is the idle mixture adjusting screw, sometimes called the CO screw. This sits horizontally at the bottom of the carburettor body under the automatic choke.

Now it's time to take off the two diaphragms, one each side of the body. The one with a lever under it is the accelerator pump, and the one which connects to the base of the carburettor via a rubber pipe is the anti-stall device. You get new diaphragms in the kit, but remember which way round they fit.

Last to be dismantled is the automatic choke. Start by undoing the three screws which hold the ring which clamps the thermostat housing. You won't be able to get at one of them with a screwdriver because the water pipe is in the way, but the screws are hexagon headed so you can loosen the last one with a small spanner.

Once they're loose you can rotate the housing and, before you take it off, identify the two scribed lines at the top which make sure you get it back in the correct position. As you lift the housing off you'll see that the coiled spring inside fits over a brass peg. Make sure it fits there when you put it back.

Under the housing is a plastic heat shield, and under that is the horizontal choke diaphragm rod. Undo the three screws at the end of the choke body and

▶ If the power valve diaphragm is bad, like this one, get a new one from your local Weber dealer. It isn't included in the overhaul kit.

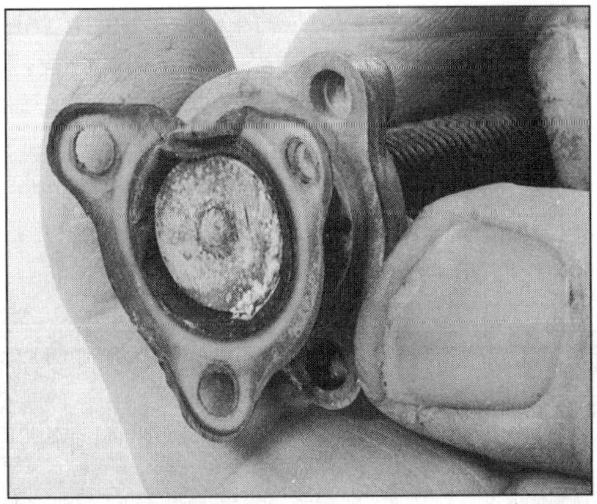

▶ You used not to get a new fuel filter in the kit, but Weber say you should now. Check the kit in case it's old stock.

▶ On the top of the body are the primary and secondary air correction jets with emulsion tubes underneath. You may have to use a piece of wood to get the tubes out. Don't use wire or you could damage the calibration.

▶ The accelerator pump jet holds down a small Y-shaped piece which sends a jet to each choke.

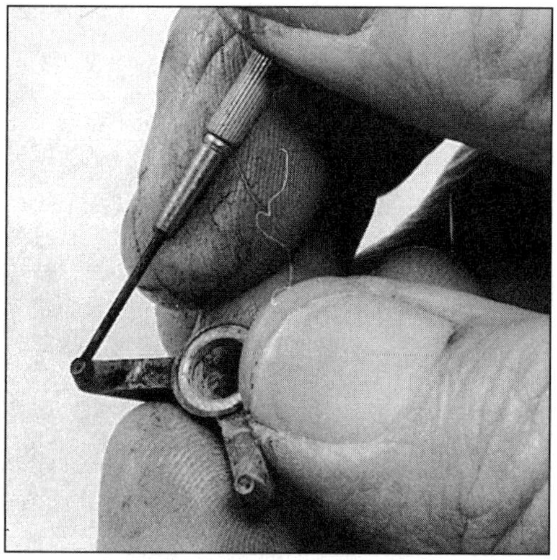

◀ *There are tiny holes in the ends of the accelerator pump Y-piece. They often get clogged, but they must be clean.*

▼ *There are three jets down in the float bowl. The two at an angle are the primary and secondary main jets. In front of them is the full-power needle valve.*

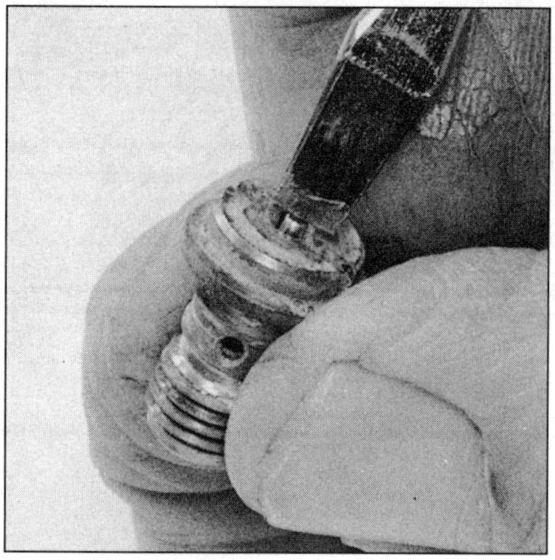

◀ *Check the needle plunger in the full-power needle valve. It mustn't be sticky or jammed.*

▼ *The two brass screws, one on either side of the body, house the primary and secondary idle jets.*

▶ *The idle jets are in two pieces, the jet itself and the holder. Take them apart for cleaning.*

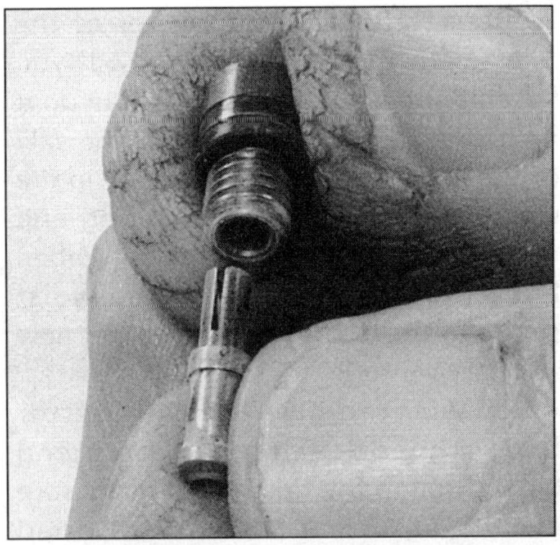

▶ *The idle mixture screw, or CO screw, is in an angled boss on the base flange.*

▼ *Look at the muck inside this accelerator pump diaphragm. You get a new diaphragm in the kit.*

▶ *This is the brass blanking plug for the accelerator pump.*

have a look at the choke diaphragm. You don't get a new one in the kit, but if it's at all suspect get a new rod and diaphragm from your Weber agent. It's held inside by a small circlip, a spring and a washer. Make a note of the order in which they come off.

At the end of the diaphragm cover plate is a small central adjusting screw under a plastic plug. Leave it as it is for the moment, you may not have to adjust it. The plug usually gets damaged when you dig it out and it's another part you don't get in the kit. If you do have to dig it out to adjust the choke setting later on, you'll need a new one.

If you want to take the automatic choke right off, as you might do if you want to change it for a manual one, there are three more screws to undo. After they're undone you can either twist it to disengage the connecting link to the throttle lever or take the small split pin out of the other end of the link.

Unless you want to change the main venturi sizes, which you won't unless you're playing about with advanced power tuning, that's all you need to take off. There's no need to take off the various links and levers and, unless they're seized, there's no need to take out the throttle or choke spindles. Open the primary throttle a few times to make sure that the secondary throttle follows it easily without sticking, and that both throttles snap shut when you let go. If they don't, a good soaking in cleaner, or a squirt or two of WD 40, will usually free them off.

Before you leave the body, check the security of the two auxiliary venturis, the 'bridge pieces' which fit in the top of the main venturis. They should be tight, but if one is loose it can cause uneven running and is a very tricky fault to spot later on. Tightening them isn't easy, but we've done it in the past by lightly knurling the side edges of the lugs which fit in the carb body. If you haven't got a knurling tool you can do it by holding a sharp file against the side of the lug and squeezing it gently in the vice. It isn't the best engineering and certainly not recommended by Weber, but it sometimes works. If the worst comes to the worst you have to get new auxiliary venturis.

Now you can give the carburettor body a good soaking and scrub and blow through the drillings to make sure everything's clean and free of dried petrol deposits. This is where the aerosol of proprietary cleaner comes in very useful. It's surprising how much muck it shifts out of the drillings.

When you put back the idle mixture screw, screw it home and then undo it three full turns. This will give you a basic setting to get the engine started. There's no need to take the centre screw out of the domed cover on the automatic choke, which carries the coolant pipe stubs, unless it is leaking. There's nothing under it that you need to get to. What's more, the screw often seizes very tight in the housing, and there's a risk of shearing it off.

Put all the parts back together again in pretty well the reverse order that you took things apart. Then you're ready to set the choke plate pull-down and the choke phasing. However, before you put the top back on the body you have to check the float level. There are two settings. For the first one you hold the top vertically so that the float is just resting on the needle valve, and measure the distance between the machined face of the carburettor top and the bottom of the float. If it isn't right, you bend the small brass tag on the float, the one which fits in the little handle on top of the needle valve.

For the second setting, the Weber manual tells you to hold the whole

▶ *Opposite the accelerator pump diaphragm there is sometimes a similar one, an anti-stall device. It isn't fitted to all DGAV models. Don't get the diaphragms in the kit mixed up.*

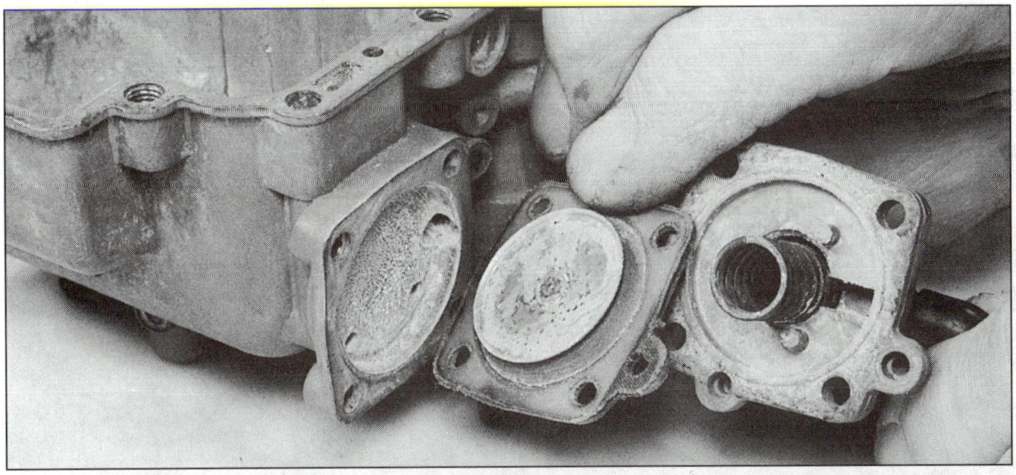

▶ *Before you take the automatic choke off, identify the two scribed marks for lining it up again.*

▶▶ *The spring in the automatic choke hooks over a brass peg. Make sure you hook it back properly when you reassemble.*

▶ *Under the automatic choke there's a plastic heat shield. Save it and clean it, you don't get a new one in the kit.*

◀ *There's also a diaphragm in the end of the automatic choke body. If it's damaged you have to get a complete diaphragm and rod assembly.*

◀ *The screw for adjusting the choke plate pull-down is in the end of the diaphragm cover under a small plastic plug.*

◀◀ *Before you leave the body, check the security of the two venturi bridges.*

◀ *There are two settings for the float. You measure the first with the float hanging vertically down.*

▶▶ The book says check the float drop with the float hanging down, but I find it easier to do with the carburettor top upside down on the bench.

▶ If either of the float settings is wrong you adjust them by bending small tabs where the float hinges.

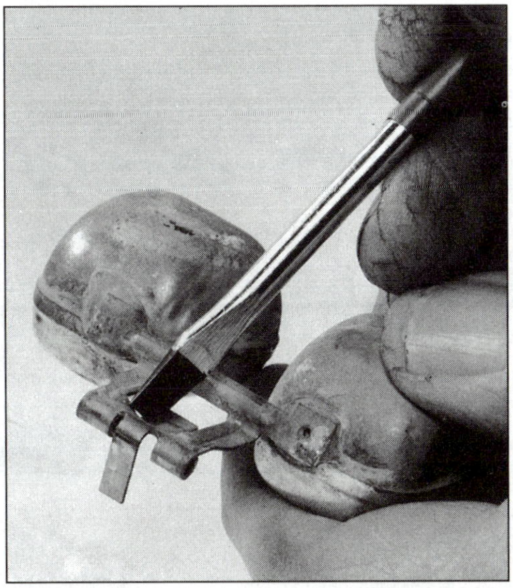

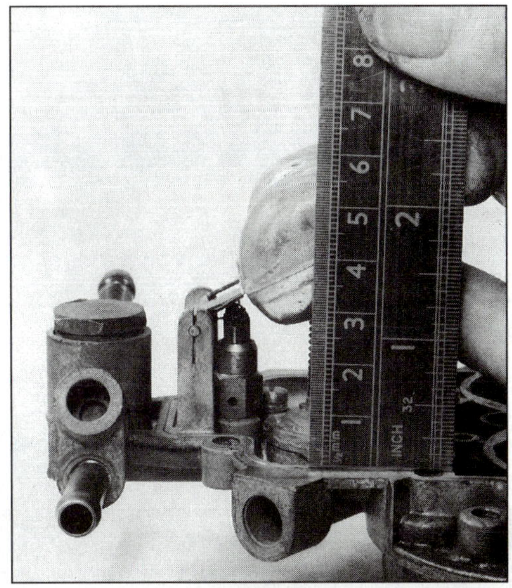

▶ To set the choke plate pull-down you need a special tool which isn't available from Weber – an ordinary rubber band.

▼ You adjust the choke phasing by bending this steel tag. Crude, but effective.

▲ *You'll find the fast-idle adjustment screw behind the choke housing.*

◀ *The screw with the large milled head is for setting the slow-idle speed.*

▼ *If you want to replace the automatic choke with a manual one, it comes off after you undo these three screws.*

thing upside down and let the float hang, but I find it's difficult to measure things upside down, so I usually lay it on the bench with the float uppermost and lift the float gently until it stops. Then you again measure the distance from the mating flange to the bottom of the float. This time, if it isn't right, you bend the small tag on the float which stops it coming down too far by hitting the needle valve housing. The float settings vary slightly with different engines but they're in the car's workshop manual.

To set the choke plate pull-down you need a special tool which Weber doesn't supply, but most stationery shops do – an ordinary rubber band. Before you refit the choke thermostat housing, hook a rubber band over the choke plate lever – the small brass pin into which the coiled spring fits – and hook the rubber band so that the chokes are closed.

Now open the throttle so that the choke plates fully close, and then let it go. The fast idle screw should now be sitting on the top step of the fast idle cam. Use a small screwdriver to push the choke diaphragm rod outwards against its stop, and measure the gap between the bottom edges of the choke plates and the walls of the air intakes. You'll find the correct gap specified in the car's workshop manual, and you measure the gap with a drill shank of the correct diameter. If it isn't correct, you have to dig out the plug from the choke diaphragm cover to get at the adjusting screw.

To set the choke phasing, partly open the throttle so that the fast idle screw is on the lower step of the cam, and then let the throttle go. Push the choke plates down until the fast idle screw jams up against the step in the fast idle cam, then let them go. Once more, check the clearance at the bottom edge of the choke plates using a drill shank. Again, you can get the correct clearance from the workshop manual.

There isn't a screw to adjust if the clearance is wrong. It's adjusted by bending the tag on the choke spindle lever with a pair of pliers – brutal, but effective.

If you can't get hold of a workshop manual for the appropriate settings, your Weber agent will give you all the figures you need if you tell him on which engine you are using the carburettor.

Some people make quite a thing about lubricating the throttle and choke spindles where they run in the body. The theory is that dunking the body in a bath of cleaner will have washed out any lubricant that was there. Possibly it will have done, if there was any there to start with. I have never bothered, and in a good many years of stripping and cleaning carburettors – from vintage to modern – I've never had problems with throttle or choke spindles sticking. But, if you feel safer with a spot of lubricant, by all means dribble some penetrating oil around the ends, or squirt on a spot of aerosol grease. Where I do use lubricant, and usually aerosol grease, is on the throttle linkages generally. I find it makes operation of them much smoother.

And that's it. It sounds very complicated, but it's all logical and straightforward when you come to do it. After fitting the carburettor on the engine, set the slow-running, fast-idle and CO mixture in the usual way.

Useful information

Specialists and suppliers

Autocar Equipment Ltd	77–85 Newington Causeway, London SE1 6BJ. Tel: 071 403 5959 *Lumenition ignition equipment*
Burton Performance Centre Ltd	623 Eastern Avenue, Ilford, Essex IG2 6PN. Tel: 081 554 2281 *Engine tuning and rebuilding/machining/performance components*
Induction Technology Group Ltd	Unit 5 Fairfield Court, Wheeler Road, Seven Stars Industrial Estate, Whitley, Coventry CV3 4LJ Tel: 0203 305386. *Free-flow air filters*
Janspeed Engineering Ltd	Castle Road, Salisbury, Wiltshire SP1 3SQ. Tel: 0722 21833 *Performance exhausts and tuning components*
K&N Europe Ltd	John Street, Warrington, WA2 7UB. Tel: 0925 636950 *Free-flow air filters*
Kent Performance Cams Ltd	1–4 Military Road, Shorncliffe Industrial Estate, Folkestone, Kent CT20 3SP. Tel: 0303 48666 *Performance camshafts and associated components*
Oselli Engineering Ltd	Ferry Hinksey Road, Oxford, OX2 0BY *Engine tuning/engine rebuilding/machining*
Piper RS Ltd	2 St John's Court, Ashford Business Park, Ashford, Kent TN24 0SJ. Tel: 0233 500300 *Performance camshafts and associated components/performance exhausts*
Specialised Engines Ltd	15 Curzon Drive, Grays, Essex RM17 6BG. Tel: 0375 378606 *Ford engine specialists*
Weber Concessionaires Ltd	Dolphin Road, Sunbury-on-Thames, Middlesex TW16 7HE. Tel: 0932 764833 *Weber carburettors*
Vulcan Engineering Ltd	185 Uxbridge Road, Hanwell, London W7 3TH. Tel: 081 579 3202 *Engine rebuilding and tuning*

Magazines

(Obtainable from all good newsagents.)
Cars & Car Conversions, Fast Car, Fast Ford, Performance Ford, Which Kit?

Books

(Out of print books can sometimes be obtained through specialist motoring booksellers and at autojumbles.)
Camshaft Tuning Manual (Kent Performance Cams Ltd)
Piper Tuning Manual by Bob Gayler (Haynes Publishing) (out of print)
Guide to the Purchase & Restoration of Escort and Cortina Mk I and Mk II by Kim Henson (Haynes Publishing)
Ford Motor Company Workshop Manuals for the following cars:
Mk I Escort (out of print)
Mk II Escort (a few copies still available from Ford dealers at the time of writing)
Cortina 1975 on (out of print)
Mk I and Mk II Capri (out of print)

Appendix

Specifications & Data for Kent Crossflow 711M engines

Engine

	All 1.1 litre engine variants
Camshaft position	In crankcase on right
Valve control	By rockers and pushrods
Firing order	1 - 2 - 4 - 3
Bore	80.98 mm (3.188 in)
Stroke	53.29 mm (2.098 in)
Swept volume	1071 cc
Idling speed	800 ± 25 rpm
Maximum continuous speed	5800 rpm

	1.1 LC	1.1 HC	1.1 2V
Engine coding	G1B	G2C	G3A
Compression ratio	8.0 : 1	9.0 : 1	9.0 : 1
Compression at starter speed	9 to 11 kp.cm^2	10 to 12 kp.cm^2	10 to 12 kp.cm^2
	(128 to 156 lbf.in^2)	(142 to 170 lbf.in^2)	(142 to 170 lbf.in^2)
Mean working pressure	8.23 kp.cm^2	8.58 kp.cm^2	8.58 kp.cm^2
	(117 lbf.in^2)	(122 lbf.in^2)	(122 lbf.in^2)
Engine output (DIN)	32 kW (44 HP)	35 kW (48 HP)	42 kW (57 HP)
	at 5500 rpm	at 5500 rpm	at 5500 rpm
Torque (DIN)	71 Nm (7.2 m.kp)	74 Nm (7.5 m.kp)	77 Nm (7.7 m.kp)
	at 3000 rpm	at 3000 rpm	at 4000 rpm

Cylinder Block

	All 1.1 litre engine variants
Cast cylinder block marking	711M-6015-A-A
Number of main bearings	5
Cylinder liner bore	84.112 to 84.175mm (3.311 to 3.314 in)
Cylinder bore diameter – Standard – A	80.947 to 80.957mm (3.1869 to 3.1873 in)
– B	80.957 to 80.967mm (3.1873 to 3.1877 in)
– C	80.967 to 80.977mm (3.1877 to 3.1880 in)
– D	80.977 to 80.987mm (3.1880 to 3.1885 in)
– E	80.987 to 80.997mm (3.1885 to 3.1889 in)
– F	80.997 to 81.007mm (3.1889 to 3.1893 in)
Bearing width	26.822 to 26.873mm (1.056 to 1.058 in)
Fitted main bearing shells, vertical interior diameter	54.013 to 54.044mm (2.126 to 2.128 in)
Undersize – 0.254mm (0.010 in)	53.759 to 53.790mm (2.116 to 2.118 in)
– 0.508mm (0.020 in)	53.505 to 53.536mm (2.106 to 2.108 in)
– 0.762mm (0.030 in)	53.251 to 53.282mm (2.096 to 2.098 in)
Main bearing bore (in block) – Standard	57.683 to 57.696mm (2.271 to 2.272 in)
– Oversize	58.064 to 58.077mm (2.286 to 2.287 in)
Camshaft bearing bore (in block) – Standard	42.888 to 42.913mm (1.689 to 1.690)
– Oversize	+0.508mm (+0.020 in)

Crankshaft	Main journal diameter – Standard	53.983 to 54.003mm (2.125 to 2.126 in)
	Undersize – 0.254mm (0.010 in)	53.729 to 53.749mm (2.115 to 2.116 in)
	– 0.508mm (0.020 in)	53.475 to 53.495mm (2.105 to 2.106 in)
	– 0.762mm (0.030 in)	53.221 to 53.241mm (2.095 to 2.096 in)
	Crankshaft end float	0.075 to 0.280mm (0.003 to 0.011 in)
	Length of main bearing shell	25.273 to 25.527mm (0.995 to 1.005 in)
	Play in main journal bearing shell	0.010 to 0.061mm (0.0004 to 0.0024 in)
	Crank pin diameter – Standard	49.195 to 49.215mm (1.937 to 1.938 in)
	Undersize – 0.05mm (0.002 in)	49.144 to 49.164mm (1.935 to 1.936 in)
	– 0.25mm (0.010 in)	48.941 to 48.961mm (1.927 to 1.928 in)
	– 0.51mm (0.020 in)	48.687 to 48.707mm (1.917 to 1.918 in)
	– 0.76mm (0.030 in)	48.433 to 48.453mm (1.907 to 1.908 in)
	– 1.02mm (0.040 in)	48.179 to 48.199mm (1.897 to 1.898 in)
Camshaft	Coding (paint ring) – 1.1 LC.	Yellow
	– 1.1 HC	Yellow
	– 1.1 2V	Red/white
	Drive	By chain with tensioner
	Thickness of camshaft retaining plate	4.470 to 4.520mm (0.176 to 0.178 in)
	Cam lift – Inlet	5.985 to 5.866mm (0.236 to 0.231 in)
	– Exhaust	5.894mm (0.232 in)
	Length of cams (between heel & tip) – Inlet	33.417 to 33.298mm (1.316 to 1.311 in)
	– Exhaust	33.326mm (1.312 in)
	Camshaft bearing diameter – Front, centre, rear	39.616 to 39.637mm (1.560 to 1.561 in)
	Bearing bush internal diameter – Front, centre, rear	39.662 to 39.675mm (1.561 to 1.562 in)
	Camshaft end float	0.06 to 0.2mm (0.002 to 0.008 in)
Piston	Piston diameter – Standard – Grade E	80.954 to 80.964mm (3.187 to 3.1875 in)
	– Grade F	80.964 to 80.974mm (3.1875 to 3.188 in)
	Piston diameter – Oversize	0.064mm (0.0025 in)
	Grade E	81.018 to 81.028mm (3.1896 to 3.190 in)
	Grade F	81.028 to 81.038mm (3.190 to 3.1904 in)
	Piston to bore clearance	0.023 to 0.043mm (0.0009 to 0.0016 in)
	Ring gap (fitted in bore) – Top	0.23 to 0.36mm (0.009 to 0.014 in)
	– Centre	0.23 to 0.36mm (0.009 to 0.014 in)
	– Bottom	0.23 to 0.36mm (0.009 to 0.014 in)
Gudgeon Pin	Length of gudgeon pin	70.99 to 71.37mm (2.795 to 2.810 in)
	Pin diameter – 1	20.622 to 20.625mm (0.8119 to 0.8120 in)
	– 2	20.625 to 20.627mm (0.8120 to 0.8121 in)
	– 3	20.627 to 20.630mm (0.8121 to 0.8122 in)
	– 4	20.630 to 20.632mm (0.8122 to 0.8123 in)
	Pin interference in piston at 21°C	0.003 to 0.008mm (0.0001 to 0.0003 in)
	Clearance in connecting rod at 21°C	0.004 to 0.010mm (0.0002 to 0.0004 in)

Connecting Rod

Bore diameter of big-end	52.89 to 52.91mm (2.082 to 2.083 in)
Bore diameter of small-end – White	20.629 to 20.632mm (0.8122 to 0.8123 in)
– Red	20.632 to 20.634mm (0.8123 to 0.8124 in)
– Yellow	20.634 to 20.637mm (0.8124 to 0.8125 in)
– Blue	20.637 to 20.640mm (0.8125 to 0.8126 in)
Vertical Internal diameter – Standard	49.221 to 49.260mm (1.938 to 1.939 in)
Undersize – 0.051mm (0.002 in)	49.170 to 49.208mm (1.936 to 1.937 in)
– 0.254mm (0.010 in)	48.967 to 49.005mm (1.928 to 1.929 in)
– 0.508mm (0.020 in)	48.713 to 48.751mm (1.918 to 1.919 in)
– 0.762mm (0.030 in)	48.491 to 48.592mm (1.909 to 1.913 in)
– 1.016mm (0.040 in)	48.205 to 48.243mm (1.898 to 1.899 in)
Clearance – Big-end journal to bearing	0.006 to 0.064mm (0.0002 to 0.0025 in)

Cylinder Head

Cast marking on cylinder head	33
Valve seat angle in head	44°30' to 45°
Stem bore, inlet and exhaust valves	7.907 to 7.937mm (0.311 to 0.312 in)
Bore for bushes	11.133 to 11.153mm (0.438 to 0.439 in)

Valves

	1.1 LC	**1.1 HC**	**1.1 2V**
Valve clearance – Inlet	0.20mm (0.008 in)	0.20mm (0.008 in)	0.25mm (0.010 in)
– Exhaust	0.55 mm (0.022 in)	0.55mm (0.022 in)	0.55mm (0.022 in)
Inlet valve – Opens	21° BTDC	21° BTDC	29° BTDC
– Closes	55° ABDC	55° ABDC	63° ABDC
Exhaust valve – Opens	70° BBDC	70° BBDC	71° BBDC
– Closes	22° ATDC	22° ATDC	21° ATDC

	All 1.1 litre engine variants
Valve springs – Number of turns	3.75 or 5.75
Cam follower diameter	13.081 to 13.094mm (0.515 to 0.516 in)
Clearance – cam follower to block	0.013 to 0.05mm (0.0005 to 0.002 in)

Inlet Valve

Length	110.668 to 111.176mm (4.357 to 4.377 in)
Valve head diameter	35.94 to 36.19mm (1.415 to 1.425 in)
Valve stem diameter – Standard	7.868 to 7.886mm (0.3098 to 0.3105 in)
Oversize – 0.076mm (0.003 in)	7.945 to 7.962mm (0.3127 to 0.3135 in)
– 0.381mm (0.015 in)	8.249 to 8.267mm (0.3247 to 0.3255 in)
Valve stem play in guide	0.02 to 0.068mm (0.0008 to 0.0027 in)
Valve stroke – 1.1 LC	8.81mm (0.347 in)
– 1.1 HC	8.81mm (0.347 in)
– 1.1 2V	8.63mm (0.340 in)

Exhaust Valve

Length	110.363 to 110.871mm (4.345 to 4.365 in)
Valve cup diameter	31.34 to 31.59mm (1.234 to 1.244 in)
Valve stem diameter – Standard	7.846 to 7.863mm (0.3089 to 0.3096 in)
Oversize – 0.076mm (0.003 in)	7.922 to 7.939mm (0.3119 to 0.3126 in)
– 0.381mm (0.015 in)	8.227 to 8.243mm (0.3239 to 0.3245 in)
Valve stem play in guide	0.043 to 0.091mm (0.0017 to 0.0036 in)
Valve stroke	8.67mm (0.341 in)

Engine Lubrication

Oil type	HD oil
Viscosity – under –12°C	SAE 5W/20
– under 0°C	SAE 5W/30
– –23°C to +32°C	SAE 10W/30, SAE 10W/40 or SAE 10W/50
– Over –12°C	SAE 20W/40 or SAE 20W/50
Ford specification	SS-M2C-9001AA
Initial capacity with filter	3.67 litres (6.5 pints)
Oil change without filter change	2.75 litres (4.8 pints)
Oil change with filter change	3.25 litres (5.7 pints)
Minimum oil pressure at – 700 rpm and 80°C	0.6 kp/cm2 (8.5 lbf/in2)
– 2000 rpm and 80°C	1.5 kp/cm2 (21 lbf/in2)
Oil pressure warning light comes on at	0.4 ± 0.1 kp/cm2 (6 ± 1.5 lbf/in2)
Excess pressure valve opens at	2.46 to 2.81 kp/cm2 (35 to 40 lbf/in2)
Oil pump play in outer rotor casing	0.1397 to 0.2667mm (0.0055 to 0.0105 in)
Gap between internal and outer rotor	0.0508 to 0.1270mm (0.002 to 0.005 in)
Axial play of outer and internal rotor in relation to oil pump cover	0.0254 to 0.0635mm (0.001 to 0.0025 in)

Tightening Torque Values

	All 1.1 litre engine variants		
	NM	**kp.m**	**lbf.ft**
Main bearing cover	75 to 82	7.5 to 8.2	55 to 60
Connecting rod bolts	42 to 48	4.2 to 4.8	31 to 35
Crankshaft belt pulley	33 to 38	3.3 to 3.8	24 to 28
Camshaft chain sprocket	17 to 21	1.7 to 2.1	13 to 15
Rear sealing ring carrier	17 to 21	1.7 to 2.1	13 to 15
Flywheel	68 to 76	6.8 to 7.6	50 to 56
Clutch thrust plate to flywheel	17 to 21	1.7 to 2.1	13 to 15
Front crankcase cover	7 to 10	0.7 to 1.0	5 to 7
Oil pump	17 to 21	1.7 to 2.1	13 to 15
Inlet line	17 to 21	1.7 to 2.1	13 to 15
Oil pump cover	7 to 10	0.7 to 1.0	5 to 7
Rocker Shaft	24 to 30	2.4 to 3.0	18 to 22
Cylinder head (1)	7	0.7	5
(2)	28 to 42	2.8 to 4.2	21 to 31
(3)	70 to 76	7.0 to 7.6	52 to 56
After 10 to 20 minutes (4)	90 to 97	9.0 to 9.7	66 to 71
After engine has warmed up (15 mins at 1000 rpm) tighten up (5)	90 to 97	9.0 to 9.7	66 to 71
Cylinder head cover	3.5 to 5	0.35 to 0.5	3 to 4
Sump (1)	4 to 7	0.4 to 0.7	3 to 5
(2)	8 to 11	0.8 to 1.1	6 to 8
Oil drain screw	27 to 34	2.7 to 3.4	20 to 25
Oil pressure switch	13 to 15	1.3 to 1.5	10 to 11
Spark plugs	30 to 39	3.0 to 3.9	22 to 29
Inlet manifold	17 to 21	1.7 to 2.1	13 to 15
Exhaust manifold	21 to 25	2.1 to 2.5	15 to 18
Fuel pump	16.3 to 20.3	1.63 to 2.03	12 to 15
Water pump	7 to 10	0.7 to 1.0	5 to 7
Thermostat housing	17 to 21	1.7 to 2.1	13 to 15
Fan to water pump flange	7 to 10	0.7 to 1.0	5 to 7
Timing chain tensioner	7 to 10	0.7 to 1.0	5 to 7

Engine

	All 1.3 and 1.6 litre engine variants
Position of camshaft	In crankcase on right-hand side
Valve control	By pushrods and rocker levers
Firing order	1 - 2 - 4 - 3
Bore	80.98mm (3.188 in)

	1.3 LC	1.3 HC	1.3 2V	1.6 2V
Engine coding	J1F	J2H	J3D	L3A
Stroke	62.99mm	62.99mm	62.99mm	77.62mm
	(2.478 in)	(2.478 in)	(2.478 in)	(3.056 in)
Swept volume	1263 cc	1263 cc	1263 cc	1588 cc
Compression ratio	8.0 : 1	9.0 : 1	9.0 : 1	9.0 : 1
Pressure at	9 to 11 kp.cm^2	10 to 12 kp.cm^2	10 to 12 kp.cm^2	11 to 13 kp.cm^2
starter speed	(128 to 156 lbf.in^2)	(142 to 170 lbf.in^2)	(142 to 170 lbf.in^2)	(156 to 184 lbf.in^2)
Mean working	8.43 kp.cm^2	9.0 kp.cm^2	9.0 kp.cm^2	10.0 kp.cm^2
pressure	(120 lbf.in^2)	(128 lbf.in^2)	(128 lbf.in^2)	(142 lbf.in^2)
Idling speed	800 ± 25 rpm	800 ± 25 rpm	800 ± 25 rpm	800 ± 25 rpm
Max continuous speed	5800 rpm	5800 rpm	5800 rpm	6000 rpm
Engine output (DIN)	40 kW (54 HP)	42 kW (57 HP)	51 kW (70 HP)	62 kW (84 HP)
	at 5500 rpm	at 5500 rpm	at 5500 rpm	at 5500 rpm
Torque (DIN)	85 Nm (8.7 m.kp)	91 Nm (9.3 m.kp)	92 Nm (9.4 m.kp)	125 Nm (12.7 m.kp)
	at 3000 rpm	at 3000 rpm	at 4000 rpm	at 4000 rpm

Cylinder Block

	All 1.3 and 1.6 litre engine variants
Cast marking on cylinder block – 1.3 litre	711M-6015-A-A
– 1.6 litre	711M-6015-B-A
Number of main bearings	5
Cylinder liner bore	84.112 to 84.175mm (3.311 to 3.314 in)
Cylinder bore diameter – Standard – A	80.947 to 80.957mm (3.1869 to 3.1873 in)
– B	80.957 to 80.967mm (3.1873 to 3.1877 in)
– C	80.967 to 80.977mm (3.1877 to 3.1881 in)
– D	80.977 to 80.987mm (3.1881 to 3.1885 in)
– E	80.987 to 80.997mm (3.1885 to 3.1889 in)
– F	80.997 to 81.007mm (3.1889 to 3.1893 in)
Bearing width	26.822 to 26.873mm (1.056 to 1.058 in)
Fitted main bearing shells, vertical internal diameter	54.013 to 54.044mm (2.126 to 2.128 in)
Undersize – 0.254mm (0.010 in)	53.759 to 53.790mm (2.116 to 2.118 in)
– 0.508mm (0.020 in)	53.505 to 53.536mm (2.106 to 2.108 in)
– 0.762mm (0.030 in)	53.251 to 53.282mm (2.096 to 2.098 in)
Main bearing bore (in block) – Standard	57.683 to 57.696mm (2.271 to 2.2715 in)
– Oversize	58.064 to 58.077mm (2.286 to 2.2865 in)
Camshaft bearing bore (in block) – Standard	42.888 to 42.913mm (1.6885 to 1.6894 in)
– Oversize	+0.508mm (+0.020 in)

Crankshaft

Main bearing journal diameter – Standard	53.983 to 54.003mm (2.125 to 2.126 in)
Undersize – 0.254mm (0.010 in)	53.729 to 53.749mm (2.115 to 2.116 in)
– 0.508mm (0.020 in)	53.475 to 53.495mm (2.105 to 2.106 in)
– 0.762mm (0.030 in)	53.221 to 53.241mm (2.095 to 2.096 in)
Crankshaft end float	0.075 to 0.280mm (0.003 to 0.011 in)
Length of main bearing shell	25.273 to 25.527mm (0.995 to 1.005 in)
Play in main journal bearing shell	0.010 to 0.061mm (0.0004 to 0.0024 in)

Crankshaft (continued)	Crank pin diameter– Standard	49.195 to 49.215mm (1.937 to 1.938 in)	
	Undersize– 0.05mm (0.002 in)	49.144 to 49.164mm (1.935 to 1.936 in)	
	– 0.25mm (0.010 in)	48.941 to 48.961mm (1.927 to 1.928 in)	
	– 0.51mm (0.020 in)	48.687 to 48.707mm (1.917 to 1.918 in)	
	– 0.76mm (0.030 in)	48.433 to 48.453mm (1.907 to 1.908 in)	
	– 1.02mm (0.040 in)	48.179 to 48.199mm (1.897 to 1.898 in)	
Camshaft	Coding (paint ring) – 1.3 LC	Yellow	
	– 1.3 HC	Yellow	
	– 1.3 2V	Red/white	
	– 1.6 2V	Red	
		1.3 litre	**1.6 litre**
	Thickness of camshaft retaining plate	4.470mm (0.1760 in)	4.520mm (0.178 in)
	Cam lift – Inlet	5.985mm (0.236 in)	5.865mm (0.231 in)
	– Exhaust	5.894mm (0.232 in)	5.895mm (0.2321 in)
	Length of cams (between heel & tip) – Inlet	33.087mm (1.303 in)	33.357mm (1.313 in)
	– Exhaust.	33.326mm (1.312 in)	33.267mm (1.310 in)
		All 1.3 and 1.6 litre engine variants	
	Drive	By chain with tensioning device	
	Camshaft bearing diameter – Front, centre, rear	39.616 to 39.637mm (1.560 to 1.561 in)	
	Internal diameter of bearing bush – Front, centre, rear	39.662 to 39.675mm (1.561 to 1.562 in)	
	Camshaft end float	0.06 to 0.2mm (0.002 to 0.008 in)	
Piston	Piston diameter – Standard – Grade E	80.954 to 80.964mm (3.1872 to 3.1876 in)	
	– Grade F	80.964 to 80.974mm (3.1876 to 3.1880 in)	
	Piston diameter – Oversize	0.064mm (0.003 in)	
	Grade E	81.018 to 81.028mm (3.1890 to 3.1901 in)	
	Grade F	81.028 to 81.038mm (3.1901 to 3.1904 in)	
	Piston to bore clearance	0.023 to 0.043mm (0.0009 to 0.0010 in)	
	Ring gap (fitted in bore) – Top	0.23 to 0.36mm (0.009 to 0.014 in)	
	– Centre	0.23 to 0.36mm (0.009 to 0.014 in)	
	– Bottom	0.23 to 0.36mm (0.009 to 0.014 in)	
Gudgeon Pin	Length of gudgeon pin	70.99 to 71.37mm (2.795 to 2.810 in)	
	Pin diameter – 1	20.622 to 20.625mm (0.8119 to 0.8120 in)	
	– 2	20.625 to 20.627mm (0.8120 to 0.8121 in)	
	– 3	20.627 to 20.630mm (0.8121 to 0.8122 in)	
	– 4	20.630 to 20.632mm (0.8122 to 0.8123 in)	
	Pin interference in piston at 21°C	0.003 to 0.008mm (0.0001 to 0.0003 in)	
	Clearance in connecting rod at 21°C	0.004 to 0.010mm (0.00015 to 0.0004 in)	
Connecting Rod	Bore diameter of big-end	52.89 to 52.91mm (2.0823 to 2.0831 in)	
	Bore diameter of small-end – White	20.629 to 20.632mm (0.8122 to 0.8123 in)	
	– Red	20.632 to 20.634mm (0.8123 to 0.8124 in)	
	– Yellow	20.634 to 20.637mm (0.8124 to 0.8125 in)	
	– Blue	20.637 to 20.640mm (0.8125 to 0.8126 in)	
Connecting Rod (continued)	Vertical Internal diameter – Standard	49.221 to 49.260mm (1.938 to 1.939 in)	
	Undersize – 0.051mm (0.002 in)	49.170 to 49.208mm (1.936 to 1.937 in)	

– 0.254mm (0.010 in)	48.967 to 49.005mm (1.928 to 1.929 in)	
– 0.508mm (0.020 in)	48.713 to 48.751mm (1.918 to 1.919 in)	
– 0.762mm (0.030 in)	48.491 to 48.592mm (1.909 to 1.913 in)	
– 1.016mm (0.040 in)	48.205 to 48.243mm (1.898 to 1.899 in)	
Clearance – Big-end journal to bearing	0.006 to 0.064mm (0.0002 to 0.003 in)	

Cylinder Head

Cast marking on cylinder head – 1.3 litre	33
– 1.6 litre	37
Valve seat angle in head	44°30' to 45°
Stem bore, inlet and exhaust valves	7.907 to 7.937mm (0.311 to 0.312 in)
Bore for bushes	11.133 to 11.153mm (0.438 to 0.439 in)

Valves

	1.3 LC and 1.3 HC	**1.3 2V and 1.6 2V**
Valve play – Inlet	0.20mm (0.008 in)	0.25mm (0.010 in)
– Exhaust	0.55mm (0.022 in)	0.55mm (0.022 in)
Inlet valve – Opens	21° BTDC	27° BTDC
– Closes	55° ABDC	65° ABDC
Exhaust valve – Opens	70° BBDC	65° BBDC
– Closes	22° ATDC	27° ATDC

	All 1.3 and 1.6 litre engine variants
Valve springs – Number of turns	3.75 or 5.75
Cam follower diameter	13.081 to 13.094mm (0.515 to 0.516 in)
Clearance – cam follower to block	0.013 to 0.05mm (0.0005 to 0.0019 in)

Inlet Valve

Length	110.67 to 111.67mm (4.357 to 4.396 in)
Valve head diameter – 1.3 litre	38.02 to 38.28mm (1.497 to 1.507 in)
– 1.6 litre	39.2 to 39.6mm (1.543 to 1.559 in)
Valve stem diameter – Standard	7.868 to 7.886mm (0.3097 to 0.3104 in)
Oversize – 0.076mm (0.003 in)	7.945 to 7.962mm (0.3128 to 0.3135 in)
– 0.381mm (0.015 in)	8.249 to 8.267mm (0.3248 to 0.3255 in)
Valve stem play in guide	0.02 to 0.068mm (0.0008 to 0.0027 in)
Valve stroke	8.81mm (0.347 in)

Exhaust Valve

Length	110.36 to 110.87mm (4.345 to 4.365 in)
Valve cup diameter – 1.3 litre	31.34 to 31.59mm (1.234 to 1.244 in)
– 1.6 litre	33.8 to 34.0mm (1.331 to 1.339 in)
Valve stem diameter – Standard	7.846 to 7.863mm (0.3089 to 0.3096 in)
Oversize – 0.076mm (0.003 in)	7.922 to 7.939mm (0.3119 to 0.3126 in)
– 0.381mm (0.015 in)	8.227 to 8.243mm (0.3239 to 0.3245 in)
Valve stem play in guide	0.043 to 0.091mm (0.0017 to 0.0036 in)
Valve stroke	8.67mm (0.341 in)

Engine Lubrication

Oil type	HD oil
Viscosity – under –12°C	SAE 5W/20
– under 0°C	SAE 5W/30
– –23°C to +32°C	SAE 10W/30, SAE 10W/40 or SAE 10W/50
– Over –12°C	SAE 20W/40 or SAE 20W/50
Ford specification	SS-M2C-9001AA
Initial capacity with filter	3.67 litres (6.5 pints)

Engine Lubrication (continued)

Oil change without filter change	2.75 litres (4.8 pints)
Oil change with filter change	3.25 litres (5.7 pints)

Minimum oil pressure at – 700 rpm and 80°C	0.6 kp/cm² (8.5 lbf/in²)
– 2000 rpm and 80°C	1.5 kp/cm² (21 lbf/in²)
Oil pressure warning light glows at	0.4 ± 0.1 kp/cm² (6 ± 1.5 lbf/in²)
Excess pressure valve opens at	2.46 to 2.81 kp/cm² (35 to 40 lbf/in²)
Oil pump play in outer rotor casing	0.1397 to 0.2667mm (0.0055 to 0.0105 in)
Gap internal/external rotor	0.0508 to 0.1270mm (0.002 to 0.005 in)
Axial play of outer and internal rotor in relation to oil pump cover	0.0254 to 0.0635mm (0.001 to 0.0025 in)

Tightening Torque Values

	All 1.3 and 1.6 litre engine variants		
	NM	**kp.m**	**lbf.ft**
Main bearing cover	75 to 82	7.5 to 8.2	55 to 60
Connecting rod bolts	42 to 48	4.2 to 4.8	31 to 35
Crankshaft belt pulley	33 to 38	3.3 to 3.8	24 to 28
Camshaft chain sprocket	17 to 21	1.7 to 2.1	13 to 15
Rear sealing ring carrier	17 to 21	1.7 to 2.1	13 to 15
Flywheel	68 to 76	6.8 to 7.6	50 to 56
Clutch thrust plate to flywheel	17 to 21	1.7 to 2.1	13 to 15
Front crankcase cover	7 to 10	0.7 to 1.0	5 to 7
Oil pump	17 to 21	1.7 to 2.1	13 to 15
Inlet line	17 to 21	1.7 to 2.1	13 to 15
Oil pump cover	7 to 10	0.7 to 1.0	5 to 7
Rocker Shaft	24 to 30	2.4 to 3.0	18 to 22
Cylinder head (1)	7	0.7	5
(2)	28 to 42	2.8 to 4.2	21 to 31
(3)	70 to 76	7.0 to 7.6	52 to 56
After 10 to 20 minutes (4)	90 to 97	9.0 to 9.7	66 to 71
After engine has warmed up (15 minutes at 1000 rpm) tighten up (5)	90 to 97	9.0 to 9.7	66 to 71
Cylinder head cover	3.5 to 5	0.35 to 0.5	
Sump (1)	4 to 7	0.4 to 0.7	3 to 5
(2)	8 to 11	0.8 to 1.1	6 to 8
Oil drain screw	27 to 34	2.7 to 3.4	20 to 25
Oil pressure switch	13 to 15	1.3 to 1.5	10 to 11
Spark plugs	30 to 39	3.0 to 3.9	22 to 29
Inlet manifold	17 to 21	1.7 to 2.1	13 to 15
Exhaust manifold	21 to 25	2.1 to 2.5	15 to 18
Fuel pump	16.3 to 20.3	1.63 to 2.03	12 to 15
Water pump	7 to 10	0.7 to 1.0	5 to 7
Thermostat housing	17 to 21	1.7 to 2.1	13 to 15
Fan to water pump flange	7 to 10	0.7 to 1.0	5 to 7
Timing chain tensioner	7 to 10	0.7 to 1.0	5 to 7